Rapid Intervention Teams

Greg Jakubowski
Mike Morton

edited by John Joerschke

Published by
Fire Protection Publications • Oklahoma State University
Stillwater, Oklahoma

Project Manager: Cindy Pickering
Design/Layout: Ann Moffat
Graphics: Ann Moffat and Ben Brock

Photographs courtesy of the following:
Abington Emergency Management: 7.32
India Asplundh: 4.3, 4.5, 4.7, 4.9 – 4.16, 5.2, 6.1 – 6.7, 7.4 – 7.6, 7.9, 7.10, 7.12, 7.13, 7.15, 7.16, 7.18 – 7.28,
7.31, 7.33, 8.2 – 8.20, 8.22, 8.24, 8.25, 8.27, 8.28, 9.5, 10.1, 10.2
Greg Jakubowski: 2.4, 5.8
Robert Marcus/Bryn Athyn Fire Company: 1.2, 3.2, 4.6, 4.8, 9.6, 10.3
Gary Pitcairn/Bryn Athyn Fire Company: 1.1, 1.5, 2.1, 2.3, 3.1, 9.1,
Rick Rotondo: 4.4, 5.1, 7.7, 7.8, 7.11, 7.14, 7.17,
Randy Yardumian: 1.3, 1.4, 2.2, 4.1, 4.2, 5.3, 8.21, 8.29, 9.2, 9.3, 9.4

ISBN 0-87939-194-4
Library of Congress Control Number: 2001130285
First Edition
First Printing, May 2001

Printed in the United States of America

Contents

About the Authors

Greg Jakubowski has been an active emergency responder for 22 years, certified as a Fire Officer I, Fire Instructor II, Emergency Medical Technician and Hazmat Technician, and currently serves as Deputy Fire/EMS Chief for the Bryn Athyn, Pennsylvania. Fire Company. Greg is a Senior Field Instructor for the Pennsylvania State and Montgomery County Fire Academies, and has co-authored Pennsylvania's Rapid Intervention Teams training program. Greg is also a contributing editor of FIRE RESCUE magazine, authoring the Tactics column for this magazine and its predecessors for more than ten years. Greg holds a BS in Fire Protection Engineering from the University of Maryland, and a MS in Public Safety from St. Joseph's University (Philadelphia, PA). He is a licensed professional fire protection engineer working for a major pharmaceutical company, and chairs the Society of Fire Protection Engineers' Fire Service Committee. Greg is currently an adjunct faculty member of the Philadelphia Community College, Holy Family College's Fire Science programs, and St. Joseph's University Environmental Protection and Safety Management Institute. He has taught emergency response programs throughout Pennsylvania, Florida, Virginia, Oklahoma, California, Nevada, Colorado, and France.

Mike Morton is a Lieutenant with the Bryn Athyn, Pennsylvania Fire Company. He is certified as a Fire Officer II, Fire Instructor I, Emergency Medical Technician, and Vehicle Rescue Technician. Mike was the driving force behind developing Bryn Athyn's RIT hands-on training, and has been a participant in the hands-on exercises for Working Fire Video's five-part series on Rapid Intervention Teams (August – December, 1997) In 1996 and 1997 Mike served as an associate staff member at the Fire Department Instructors Conference (FDIC) in Indianapolis where he attended classroom and practical FAST/Firefighter Survival training sessions and seminars. Together with Greg Jakubowski, he developed a 16-hour FAST training program originally presented for the Indiana Fire Instructor's Association. He has also written about RIT for Fire Rescue magazine. Mike has a BA in education/history from Bryn Athyn College, and an MA in history from Villanova University.

Acknowledgements

The authors cannot conclude their work on this book without thanking several individuals who provided us with the initial impetus to study and practice rapid intervention techniques. First, the officers and members of the Willow Grove, Pennsylvania, Fire Company who approached the Bryn Athyn, Pennsylvania, Fire Company in 1995 and asked us to learn about RITs and to be their primary rapid intervention team. Tom Sullivan, director of emergency services for Upper Moreland Township, Pennsylvania, also contributed to this as well as developing some of the initial training programs offered on this topic in Pennsylvania. Second, Dr. Jim Cline, of the New York City Fire Department and EDCON Associates. Jim provided our initial training, and much of what we originally learned came from his "Firefighter Assistance and Search Team" presentation and document. Jim's wisdom has greatly influenced us, along with many other firefighters. Third, Robert Linsinbigler, director of the Montgomery County, Pennsylvania, Fire Academy, and his staff, who supported our efforts in further developing this topic and some of the techniques utilized in it.

We are especially grateful to the officers and members of the Bryn Athyn Fire Company, particularly Chiefs R. Scott Cooper and C. Kristopher Smith, who have been so supportive in developing and implementing the RIT concept within our department.

Thank you to the librarians at the National Fire Academy in Emmitsburg, Maryland, for their assistance with locating virtually every source related to rapid intervention teams, and also thanks to the numerous authors and individuals who have selflessly dedicated their time and energy to the preparation of these materials, many of which appear in our bibliography.

Thanks also to the many people who helped us obtain the hundreds of photographs gathered for this book. Although the individual photographers are credited throughout the text, thanks must go to those who took part in the numerous photo shoots. First, thanks to the Bryn Athyn Fire Company's Thursday night duty crew, who subjected themselves to almost endless dragging and carrying, and to the other Bryn Athyn and Huntingdon Valley Fire Company members who assisted us. Second, a huge thank-you to the Lower Providence Fire Department and particularly their chief, Bryan McFarland, for their generosity in arranging a facility and providing manpower for the photography of the exterior evolutions appearing in chapter 8.

We would also like to acknowledge the work of the individuals at Fire Protection Publications for the huge effort involved in putting this work together. In particular, the project manager, Cindy Pickering; and the graphic designers, Ben Brock and Ann Moffat. We would also like to thank the editor, John Joerschke.

Finally, thanks to our families, particularly our wives Tami and India for putting up with us while we were writing this book, and to Lois Ellis Jakubowski, Greg's mother, who made sure our English was correct!

Foreword

By Lois Ellis Jakubowski

The term *fire fighting* conjures up images of village folks scurrying with water-filled buckets to help extinguish a fire in their neighbor's field, barn, or home. Later come the images of a horse-drawn pumper with bells clanging and steeds galloping furiously to the aid of their community. These sights and sounds arouse feelings of brotherhood, selflessness, and service. In the twentieth century, brightly colored trucks came to symbolize those same qualities. Today, firefighters have access to highly sophisticated equipment that is essential to address the complex challenges they will face in the twenty-first century.

Despite the changes in equipment and technology, the basic tenets of fire fighting remain constant. The primary objective is to protect the lives and property of the citizenry. The main motivation for entering the fire service is the desire to help others. Whether personnel in a fire-fighting organization are compensated or volunteer, fire fighting is a noble calling. Those who choose to answer it deserve every possible assurance that their lives are valued. This assurance is made clear when the community provides support in terms of equipment and training. When this relationship of mutual concern is in balance, its effect is reflected in a high level of performance. This relationship's logic sounds simple, but its reality poses many challenges. One of them is to organize and integrate rapid intervention teams into fire-fighting operations.

Introduction

Firefighters in colonial America showed little regard for their own safety. They recognized the importance of extinguishing a fire quickly before it could spread from home to home across the combustible roofs that were common at the time. The loss of a home could devastate a family, and everyone who could fight a fire turned out to help their neighbors. Firefighters were injured and killed in all sorts of mishaps, and they often threw caution for their own safety to the wind. Later, competing fire companies even fought each other for the chance to battle a blaze, gain notoriety, and perhaps be rewarded by the insurance company for putting first water on the fire or for quelling the flames.

As generations passed, however, career and volunteer firefighters became more professional. They began to recognize the risks they faced and to manage them better. Hazards became more complicated, and handling them required more personnel. Safety efforts blossomed in this environment, but not until the end of the twentieth century did the American fire service give the concept of rapid intervention—a plan for rescuing emergency responders in peril—the attention it deserves.

Some career fire departments formed heavy rescue companies decades ago. Their purpose was to provide specialized equipment and specially trained personnel to extricate trapped victims from a variety of situations. Although the term was essentially unheard of at the time, these units also functioned as early rapid intervention teams, manned and equipped to rescue firefighters who became trapped on the job. Over the years, incidents became more complicated, overall manning was reduced, and fire service management recognized the rescue companies' value in many situations. As they evolved, rescue companies found themselves more often involved in incidents and less able to perform their secondary duties of rapid intervention for firefighters. This problem, however, was not widely recognized until the 1990s, when responders got into trouble and rescue companies were not readily available. In other communities protected by career firefighters today, the on-duty manpower complement may be barely adequate to provide basic structural fire-fighting services. Once a working fire is identified, off-shift personnel are paged from home. Not until these off-shift personnel are all paged back will mutual aid be called. This time-consuming process places personnel operating at the incident at risk with no backup should something go wrong.

Decreased manning in many volunteer fire departments paralleled that in the career departments. From twenty to thirty or more years ago, volunteer organizations typically had large numbers of active members. They could muster significant manpower pools at working incidents. If something went wrong, the chief or the incident commander simply turned to the manpower pool and deployed two, four, six, or more personnel as needed to deal with the emergency. Today, this simply is not possible in many communities protected by volunteers. Over time, the number of active volunteers dwindled. If their departments are lucky, they have enough personnel to respond with the primary apparatus due on the alarm. Calls that these communities could handle on their own in the past now require one or more automatic aid companies. Responding units come with a handful of personnel, barely enough to accomplish the immediate job, let alone handle a sudden, unexpected event. Should something go wrong, personnel often find themselves at risk. In such circumstances, having a rapid intervention plan in place can quickly save lives.

The rapid intervention team concept involves many terms. The National Fire Protection Association (NFPA) uses the basic term *rapid intervention crew* (RIC). According to the NFPA, a rapid intervention crew should include "at least two members, and shall be available for rescue of a member or a team if the need arises. Rapid intervention crews shall be fully equipped with the appropriate protective clothing, protective equipment, SCBA, and any specialized rescue equipment that might be needed given the specifics of the operation under way."[1] The concept of the rapid intervention crew has grown into that of a rapid intervention team (RIT), which generally consists of more than two members. The New York City Fire Department and others have dubbed their rapid intervention units firefighter assist and search teams (FAST).

Among the many other names for rapid intervention teams are:

- Immediate response team (IRT)
- Rescue assist team (RAT)
- Firefighter rescue available team (FRAT)
- Rapid deployment unit (RDU)
- Firefighter assist team (FAT, not widely used, for obvious reasons)

The fire service loves acronyms as much as any other government agency.

No matter what they are called, rapid intervention teams must be an integral part of any emergency response organization. This book's intent is to provide emergency response organizations with the tools to develop or enhance their rapid intervention team programs. Throughout this book, the terms *firefighter* and *emergency responder* are used interchangeably, as are *rapid intervention team (RIT)* and *firefighter assist and search team (FAST)* and *fire department* and *emergency response organization*. Rapid intervention is necessary not only for the fire service but should be considered by any emergency response organization whose personnel enter environments that can place their lives and health in immediate danger. All emergency responders and personnel who may act as incident commanders at emergency incidents should be familiar with rapid intervention concepts so that they can effectively use this risk management tool.

Note

1. NFPA 1500, *Fire Department Occupational Safety and Health Program*, 1997 edition.

Chapter 1
Why Rapid Intervention?

Why Rapid Intervention? | **Chapter 1**

About 100 firefighters die on the job across the United States annually. These tragedies typically are reported through government and nongovernment agencies such as the National Institute for Occupational Safety and Health (NIOSH) and the National Fire Protection Association (NFPA). They demonstrate most emphatically the need for safety improvements. Effective rapid intervention teams might have saved lives in any number of recent firefighter tragedies:

1988   New Jersey. Five firefighters were killed while fighting a fire in an auto dealership. The firefighters were trapped after the roof collapsed. An important factor was the inadequate manpower available to mount an effective rescue effort after the collapse.

1995   Pennsylvania. Three firefighters were killed in a four-level house fire when a collapse occurred and they were lost. No effective accountability system was in place.

1995   California. An automatic garage door closed behind firefighters as they tried to back out their hand line during a rapidly intensifying house fire. One firefighter was killed and two seriously injured because other firefighters on the scene were not immediately prepared to mount an effective rescue effort.

1995   Kansas. One firefighter was killed and a captain suffered smoke inhalation after the firefighter fell through the floor into the basement. The firefighter attempted to protect himself from the fire with a hose line, but his self-contained breathing apparatus (SCBA) ran out of air before rescue could be made.

1998   North Carolina. Two firefighters died trying to exit a burning auto salvage storage building. During an interior attack, the crews began to exit, but an intense blast of heat and thick black smoke covered the area, forcing firefighters to the floor. A chief became disoriented and lost in the building. When a lieutenant reentered the structure to search for the chief, he, too, became disoriented and failed to exit. At least one other fire officer at the scene was pulled unconscious from the building.

1998   Mississippi. Two firefighters died of smoke inhalation while performing separate tasks during a fire at a strip mall. One firefighter was part of an interior attack, while the other was part of a roof ventilation team. Both were trapped when the roof collapsed. The air cylinder supply on the scene was exhausted, delaying further rescue efforts for more than an hour. One victim was found 15 feet inside the front doorway.

1998   Ohio. Two firefighters died of smoke inhalation while trying to exit the basement of a single-family dwelling after a back draft occurred in a suspended ceiling area. The victims' SCBAs ran out of air while they were trying to escape. Other firefighters made additional rescue attempts but were thwarted by excessive heat and smoke and the lack of an established water supply.

1999   Iowa. Three firefighters died during efforts to rescue civilians trapped in a house fire. The exact circumstances are unknown, but inadequate manpower, a nonexistent accountability system, and the lack of a rapid intervention team contributed to the deaths.

These are just a few random examples of recent fires where the absence of an effective emergency rescue plan may have contributed to firefighters' deaths.[1]

In addition to fatalities, hundreds of firefighters are injured every week. Many departments simply accept these injuries as a part of the job and sometimes do not even report them. The fire service has much more complete information on the number and circumstances of firefighter deaths than it has on firefighter injuries. Somehow, the firefighter grapevine picks up the word on deaths quickly, but injuries are a different story; therefore, the causes of fatalities are more clearly documented. To prevent dangerous situations from occurring in the first place, fire service organizations must study the causes of both injuries and fatalities and identify the problems that can be corrected *before* accidents occur.

Some of the commonly identified causes of firefighter casualties are:

- Failure to recognize rapidly deteriorating conditions

- Poor survival training

- Poor communication

- Inexperienced officers

- Failure to use safety equipment

- Water loss

- Freelancing

Recognizing these problems and taking steps to correct them can help to prevent many firefighter injuries and fatalities every year.

Failure to Recognize Rapidly Deteriorating Conditions

Fire is a powerful force of nature. It might not easily awe those who have not experienced its power, but firefighters know how quickly conditions can go downhill during an incident. Despite all of man's efforts, including protective clothing, breathing apparatus, heavy hose streams, and other technologies, fire can be unpredictable and can overcome even experienced firefighters. Novices are at still higher risk. Modern uses of energy-efficient windows and other construction methods allow fire conditions, including heat and smoke, to build rapidly. With minimal personnel and multiple assignments, firefighters may neglect ventilation, thus allowing built-up heat and smoke to intensify (Figure 1.1).

Fire-fighting crews often have difficulty identifying hazards on the fire ground simply because they are understaffed. They are busy performing multiple demanding tasks and overlook ongoing size-up. Firefighters' natural desire to attack the immediate problem and finish the job quickly can result in critical errors such as opening doors, walls, or ceilings without having a charged hose line readily available. That may be all a hidden fire needs to erupt into a deadly inferno. Or in their haste to complete the job, firefighters are likely to forget about maintaining an escape route. As they

Figure 1.1 Ventilation is critical to prevent a bad situation from getting worse.

progress, they might forget to look back and may wind up becoming trapped by rapidly changing conditions.

Poor Survival Training

Many fire departments just do not properly train their firefighters to survive when things go wrong. They focus on their customers—extinguishing their fires and rescuing any victims. Typically, only a small percentage of emergency responders knows quickly what to do when something goes wrong. Most others have to stop and think, and this precious reaction time might delay their rescue long enough to cause injury or death. Most firefighters are strong-willed, independent individuals who do not like to admit they are in trouble or lost. This attitude can only make things worse when situations go terribly amiss on an emergency scene.

Training firefighters to minimize the potential for bodily harm is essential. They must recognize when they are in trouble, know how to call for help, and understand how incident commanders and others must react to a responder in trouble. Yet, most fire departments do not even have a simple procedure for what to say when a firefighter gets into trouble—a situation where communications must be clear. Further, firefighters must be intimately familiar with their SCBA and know what to do if it malfunctions. This can only be accomplished through ongoing training and practice.

A few programs are shining examples of how to provide adequate survival training. Every firefighter should have the opportunity to participate in a Smoke Divers (Breathing Apparatus Specialist) course, firefighter survival course, or similar training. A number of reputable training organizations such as the Maryland Fire and Rescue Institute, Florida State Fire Training, the Burlington County Fire Academy in New Jersey, and the Iowa State University Fire Service Extension offer these programs. Openings in these programs are limited, however, and few line firefighters will attend such courses during their careers. Individual departments need to allocate time for survival training, but the unfortunate reality is that many provide little or no training in this area. Consequently firefighters are poorly prepared to face the worst scenarios.

Poor Communication

Communication failures are routinely cited as the reason for downfalls in all sorts of relationships. Few emergency services organizations rave about their communication systems, either technological (radio/dispatch) or organizational (meetings, coordination, notification of policies and procedures). Consequently, communication problems are almost certain at working incidents (Figure 1.2).

Figure 1.2 Multiple units on multiple frequencies cause havoc at working incidents.

Many emergency services organizations still communicate on low-band or other unreliable frequencies. Some share channels with other organizations, while in others numerous departments must share a limited number of channels, which adds to interference during storms or other major incidents. Still others have to make do with obsolete equipment that may or may not get the message through. These questionable arrangements can endanger firefighters should an incident go downhill.

Inexperienced Officers

As the number of fire calls is dropping nationwide, firefighters are becoming less experienced in their main responsibility—fighting fires. Meanwhile, calls for other services are increasing. Fire departments that provide first-responder emergency medical service (EMS) typically answer two or three EMS calls for every fire call they run. Additionally, many fire responses are for alarm system malfunctions, burned food, or other minor situations. Even career firefighters who work every two or three days may go months or years before they are in early at a working structure fire.

This inexperience may be most apparent in departmental training. Training is limited, and instead of hands-on exercises, it consists mostly of war stories about "how we did it when." Often only a few members of the department attend, and preparation for the serious incident will be inadequate. Inexperience may also be evident in a department's equipment. The equipment on the department's apparatus may not fit the hazards in the community. Or the system may not be integrated from water supply through agent application equipment that can deal with the range of incidents to which the department might respond. Size-up is the next casualty of this inexperience. Officers cannot distinguish a room and contents fire from a fire that has entered walls, ceilings, or other void spaces. This creates a potential death trap for firefighters who do not arm themselves with the appropriate weapons to battle the blaze. Additionally, firefighters do not show up ready to do the job. They fail to anticipate the tools they will need, and when they arrive at the scene they have to return to the apparatus to retrieve them.

Inexperienced officers and firefighters can commit other mistakes that contribute to their own demise. They may place hose lines incorrectly or size them too small for the fire encountered. Fire showing in the front of a building, for instance, may actually be handled better with hose line entry from the rear (Figure 1.3). Unfortunately, a first-in apparatus pulling up to the front of the building is probably best positioned to lay hose to the front of the building. Operating the hose line from this position, however, may endanger firefighters operating in the rear or searching above the fire for victims. Furthermore, inexperienced officers may not recognize when they have reached the point of no return from a building fire.

Figure 1.3 Initial hoseline placement can make or break an incident.

Figure 1.4 Liberal ladder placement provides alternate egress routes.

Inexperience is also a key factor in many departments' failure to ventilate fire buildings adequately or punctually. Untrained firefighters and officers may not recognize the importance of early ventilation to control a situation and possibly prevent injuries or other sudden, unexpected events such as flashover or back draft. With an increasing emphasis on customer service, inexperienced firefighters may not realize how important ventilation is to firefighters operating inside, and they will hesitate to open a structure, hoping to avoid seemingly unnecessary property damage. Finally, when manpower is limited, even experienced personnel may be too busy to remember this critical task.

Another crucial mistake that inexperienced or complacent departments make is to overlook laddering the fire building (Figure 1.4). The fastest routes in and out of a building are usually the built-in stairways. Unfortunately, rapid vertical egress for building occupants also usually results in a path for rapid vertical fire spread. Fire may quickly follow firefighters up the stairs, preventing their exit through the point of entry. Quickly and routinely laddering multiple windows creates vital escape routes. The Boston Fire Department is renowned for its liberal use of ground ladders, while the New York City Fire Department is similarly renowned for its use of aerial ladders. All fire departments could benefit from these examples, providing firefighters numerous choices for exiting a building gripped by fire.

Identifying and securing means of egress is critical as firefighters penetrate a burning occupancy. When conditions deteriorate, doors and other egress points may not be properly controlled and available to responders. Also, security equipment in buildings often prevents firefighters from quickly using these means of egress when a desperate need arises. Not only identifying the means of egress is important; clearing them of security equipment is equally essential to the safety of personnel operating inside. This skill warrants careful attention, but it is not universally taught to firefighters.

The blame for inexperience does not fall squarely on the individual. Department management must also assume major responsibility for placing people in positions of authority without providing them the appropriate training and resources to accomplish the job. What are the qualifications for department officers? What training have they completed? Is ongoing training provided? Are they given opportunities to gain experience elsewhere? All of these areas demand attention in order to minimize risks to department members.

Failure to Use Safety Equipment

Emergency responders like to present a "tough" image. Despite all of the engineering efforts put into SCBA and personal alert safety systems (PASS), firefighters still find ways to avoid using them or to avoid using them properly (Figure 1.5). Among the excuses given for not wearing SCBA into fires are that they are too heavy and cumbersome or that they interfere with the job. Other excuses might be that the incident "seemed small at first" or that "it didn't look like hazardous materials were involved" or "it slipped my mind." In the case of PASS devices, emergency response organizations try to circumvent the "I forgot" excuse by attaching a device directly to each SCBA. Activating the device, however, is just one more task to remember and can be easily forgotten in the heat of battle.

New SCBA with built-in PASS devices are designed to overcome these limitations, but they can only work if firefighters have and actually wear the equipment when they need it. Even when a strong SCBA program is in place, departments must also have a comprehensive maintenance program, including regular (daily or weekly) SCBA checkouts to ensure that they are ready for service

Figure 1.5 SCBA use at working fires is critical and should be mandatory.

at all times. SCBA can protect and even save the life of a firefighter. To do so it must work properly every time it is donned. Every fire department and every emergency responder must take responsibility for ensuring that equipment is always ready for service and is used whenever it is needed.

Water Loss

Limitations on manpower and training result in sloppiness, and in the emergency response business, sloppiness almost certainly will lead to casualties. One of the first places these limitations manifest themselves is in the failure of firefighters and officers to establish an adequate, continuous water supply at incidents. Firefighters in areas without hydrants are usually much more sensitive to this issue than firefighters in hydranted areas. But even in hydranted areas, many places create challenges to establishing an adequate, continuous water supply. Trying to control a working fire on tank water is like rolling the dice in Las Vegas. You may win some of the time, but when you lose, you can lose big time. And fire fighting is a business that cannot afford losses—the stakes are too high.

Before large diameter hose was widely utilized, most departments' primary supply lines were 2½-inch or 3-inch hose. A large percentage of engine companies carried dual beds of supply line and laid both lines at working fires to obtain the necessary water flows. This also provided a safety margin that some officers may not have considered—a backup water supply if one of the lines was lost, damaged, or otherwise interrupted. With the transition to large-diameter supply lines, many wonderful things happened. In many cases, a single line could provide all the water necessary to control the fire. It could be laid, charged, and picked up faster than dual, smaller lines. This made life easier, but eliminated the safety of a backup water supply at working incidents. Should the single supply line burst or be run over by a vehicle, or should the hydrant or water supply fail, all water supply to the fire ground could be cut off.

A number of tactics can minimize the impact of having no backup supply line. These include requiring that all vehicle water tanks are topped off once a hydrant line is charged at a scene. But nothing beats dropping a second line. In addition to the safety factor, fire officers may be pleasantly surprised by the flow that a single engine supplied by multiple large-diameter lines can produce.

Freelancing

Current documentation may not indicate the significance of freelancing in firefighter deaths. Firefighters who die in the line of duty are considered heroes and are rightly memorialized for giving their all to their community. When a firefighter dies, however, the fire department may be reluctant to admit that freelancing contributed to the fatality.

With limited manning and with reduced and inexperienced supervision, freelancing is likely to flourish. In some cases, freelancing may be the fault of the individual firefighter, but in other instances, fire department supervision may be to blame. Firefighters are typically resourceful, ambitious, and capable of making independent decisions when needed. Without a strong command structure, firefighters will take whatever actions they see fit to address the incident at hand. That is what they train for and what they live for. If these actions do not fit a carefully reasoned, well organized action plan, they essentially amount to freelancing.

Responders get away with freelancing without injury every day across the country. In turn, this may breed more freelancing and, possibly, complacency. Ultimately, however, in a sudden, unexpected event freelancers may place themselves and their colleagues in precarious situations. This becomes even more perilous when there is no effective incident commander to oversee the big picture or when incident commanders do not know where their personnel are or what they are doing.

Even in departments that use strong incident command systems, some firefighters tend to "forget" procedures in the face of a working incident. Or they may have inadequate training or practice in departmental procedures. Or if firefighters know the procedures, the organization may not regularly enforce them. These are all potentially dangerous conditions that can have serious and even fatal consequences.

Conclusion

Understanding the potential dangers discussed in this chapter is mandatory. That they cause injuries and death in the fire service is a fact. While many of these situations may not exist in every fire department, the presence of one or more of them is likely. Rapid intervention teams can provide a safety net at emergency scenes. Every firefighter should fully understand this information, particularly members of rapid intervention teams. Not only will this knowledge help RIT personnel to prepare for and overcome the emergencies they encounter, but it will also make them better, safer firefighters overall. Most important of all, these measures can prevent personal injury and death.

Note

1. These examples are taken from the following sources: *National Fire Protection Association Journal*, July/August 1995 and July/August 1996 issues; NFPA investigation report for 1999; National Institute for Occupational Safety and Health (NIOSH) August, 1999 Alert, *Preventing Injuries and Deaths of Firefighters Due to Structural Collapse*, DHHS(NIOSH) Publication No. 99-146.

Chapter 2
Sudden Unexpected Events

Sudden Unexpected Events

The firefighter's duty is to respond to the emergency calls of others. Firefighters and other emergency responders are expected to arrive promptly, rapidly assess the situation, and address the crisis at hand. They are not used to responding to their own emergency calls, and they are certainly not used to calling for help for themselves when they get into trouble. The causes of firefighters' troubles are sudden unforeseen events. These usually occur when least expected on the fire ground or emergency scene, placing responders at immediate risk. Examples of sudden unexpected events include:

- *Lost/Trapped/Unaccounted for Firefighter.* Realizing that a firefighter is missing gives fire officers one of the most sinking feelings they can have. This awareness can turn a well-managed fire ground into sudden chaos. Everyone stops what he is doing to focus on the missing firefighter. While this initial gut reaction might seem admirable, it can also allow an emergency that has been contained to flare out of control.

- *Flashover.* A flashover can completely engulf an entry team in a room or other confined area in split seconds. Flashover conditions occur when almost all combustibles in the area reach their ignition temperatures and ignite simultaneously (Figure 2.1). In these conditions, even firefighters wearing complete turnout gear and SCBA will almost certainly be burned, and they will have only seconds to escape before succumbing to the flashover. The amount of combustible material in the area will dictate if the heavy fire conditions continue unabated or subside somewhat. Quickly extending hose lines into the flashover area may be the only way to save the lives of any personnel trapped inside.

Figure 2.1 Firefighters must recognize conditions that can lead to flashover.

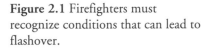

- *Back Draft.* The popular movie *Backdraft* did wonders in marketing for the fire service. It also gave the public the impression that back drafts are more common than they really are. Most firefighters have never actually seen a back draft occur. If a firefighter has witnessed a back draft, it more than likely was a relatively minor one such as occurs when opening a ceiling or other concealed space to search for hidden fire (Figure 2.2). Back drafts should not be underestimated, however, for when they do occur, they happen suddenly, with little or no warning. They are more likely when there has been a delayed alarm or in buildings that are weather tight or highly energy efficient. Double-pane windows, used frequently in today's construction, may enhance the possibility of a back draft. When a severe back draft

Figure 2.2 All firefighters must be able to recognize potential backdrafts.

occurs with responders on the scene, there will almost certainly be injuries and possibly a structural collapse.

- *Rapid Fire Increase.* A rapid fire increase typically results from either a flashover or back draft; however, fire also can rapidly attack a stairway or other void space, trapping firefighters above the fire (Figure 2.3). These situations either require an alternate means of escape such as another stairway or a ladder or require firefighters to work to regain the primary means of egress. To prevent casualties, crews likely will need to rapidly apply ground or aerial ladders and/ or advance hose lines.

- *Explosion.* The subject of hazardous materials has received continual attention in emergency services training for the past 10–15 years. Firefighters' training and preparation for dealing with exotic chemicals is better than ever before. They anticipate explosions or other serious events at locations where hazardous materials are known to be manufactured, stored, transported, or used. But in the rush to explore the unknown, more common problems may be overlooked. Gasoline, propane, and natural gas frequently are present in the built environment. When firefighters enter a burning garage, they may be concerned more about the spray can of insecticide on the shelf than about the two-gallon plastic container of gasoline on the floor. Yet the gasoline container presents a greater potential for immediate injury than does the insecticide can. Potential terrorism looms on the horizon as another job hazard. This may be particularly dangerous because terrorists often use secondary devices to drive home their point.

Figure 2.3 Firefighters working above a fire must be cognizant of rapid fire increase below.

- *Collapse.* A collapse is likely to jeopardize more than one firefighter. In the two basic collapse scenarios either a building (wall, ceiling) falls on top of responders or responders fall one or more stories through collapsed decking (floor, roof) and land in a hazardous area.

- *Cardiac Emergencies.* The nature of fire fighting demands physical fitness, yet cardiac emergencies are common in the fire service (Figure 2.4). Most firefighters do not immediately associate them with their comrades' job-related deaths, but cardiac emergencies occur frequently, suddenly, and without warning. Annual statistics consistently show that heart attacks cause 40 percent or more of firefighters' on-duty deaths. The circumstances may vary, but on any fire ground, at any incident, a firefighter may suddenly drop due to cardiac arrest. Smoke, toxic gases, and the stress of the incident all contribute to this prospect.

Are fire departments prepared for these situations? When they occur, the firefighters in the vicinity likely will once again drop whatever they are doing to assist colleagues who are in trouble. Even at incidents where a strong incident command system (ICS) is in place, such well-intended actions can completely disrupt the system, with freelancing reigning supreme. Chaos can ensue, and firefighters may place their trapped comrades at even greater risk by losing control of the incident and allowing the threat to increase. Firefighters must plan and train for these eventualities until their reactions to them are automatic. Training should include hands-on drills as well as the reviews and critiques of videos of similar events

Figure 2.4 Rapid intervention teams must be ready to handle a sudden cardiac event.

experienced by other departments. In this way, others will learn from past mistakes and avoid repeating them.

Ideally, plans must be in place to manage all types of sudden, unexpected events in order to keep the emergency scene under control. Fire departments operating under a solid ICS with firmly established rapid-intervention plans are much more likely to manage crises successfully. When there is no plan, and even sometimes when there is one, incidents tend to get out of control during and immediately after any sudden, unexpected event. Rapid intervention personnel must be familiar with any sudden, unexpected events they might have to manage. Moreover, they must be prepared with the training and equipment to take immediate and appropriate action to overcome these situations.

Chapter 3
Applicable Codes/Standards

Applicable Codes/Standards

Chapter 3

A number of regulatory, governmental, and standards-making bodies, including the National Institute for Occupational Safety and Health, the National Fire Protection Association, and the Occupational Safety and Health Administration (OSHA), have referenced rapid intervention in their publications. These organizations' recommendations may or may not apply statutorily to any particular organization, depending upon its type, the state and/or community in which it practices, and other factors. The individual fire department is responsible for determining the statutory requirements pertaining to its specific operations.[1]

NIOSH

NIOSH is a branch of the Centers for Disease Control and Prevention (CDC) in the United States Department of Health and Human Services. Since fiscal year 1998, it has conducted the Fire Fighter Fatality Investigation Project to prevent firefighter line-of-duty fatalities. The program also collects and disseminates data related to firefighter injuries and fatalities. Additionally, NIOSH had an interest in studying and preventing firefighter injuries and deaths even before fiscal year 1998. In a September 1994 alert, *Preventing Injuries and Deaths of Fire Fighters* (Publication No. 94-125), NIOSH recommended that departments should:

> Employ a buddy system whenever fire fighters wear SCBAs [Figure 3.1]. Fire fighters who wear breathing apparatus should never enter a hazardous area alone. Two fire fighters should work together and remain in contact with each other at all times. Two additional fire fighters should form a rescue team that is stationed outside the hazardous area. The rescue team should be trained and equipped

Figure 3.1 Firefighters must work in pairs to reduce risks.

to begin a rescue immediately if any of the fire fighters in the hazardous area require assistance. A dedicated rapid-response team may be required if more than a few fire fighters are in the hazardous area.

In an August 1999 alert, *Preventing Injuries and Deaths of Firefighters Due to Structural Collapse* (Publication No. 99-146), NIOSH further recommended:

Fire departments should take the following steps to minimize the risk of injury and death to fire fighters during structural fire fighting: Ensure that the incident commander conducts an initial size-up and risk assessment of the incident scene before beginning interior fire fighting. Ensure that the incident commander always maintains accountability for all personnel at a fire scene both by location and function. Establish rapid intervention crews (RICs) often called rapid intervention teams and make sure they are positioned to respond immediately to emergencies. Ensure that at least four fire fighters are on the scene before beginning interior fire fighting at a structural fire (two fire fighters inside the structure and two outside).

The same August 1999 alert also states:

The primary purpose for a RIC is to provide a dedicated and specialized team of fire fighters ready to rescue fire fighters who become trapped in a burning structure. A RIC is vitally important at a structure fire, as it provides the incident commander with a designated emergency team and thereby eliminates the need for reassigning other fire fighters to this duty during a critical period. The RIC's primary duty is to respond to emergencies in which fire fighters are trapped, lost, or disoriented in a burning structure. Under optimum conditions, a RIC should respond with the first alarm to eliminate later response time. The RIC should be equipped with full turnout gear, SCBAs, portable radios and lights, axes, forcible entry tools, hooks, and other equipment needed for the rescue effort. The RIC should report directly to the incident commander and be nearby to await rescue commands. A RIC should consist of at least two fire fighters, but the size and complexity of the incident dictates the size of the RIC.

NIOSH is not a code-making body but an agency charged with conducting research and making recommendations for the prevention of work-related disease and injury. Although NIOSH does not establish standards, it can make recommendations to OSHA to develop standards dealing with issues or problems uncovered during its research.

NFPA

The National Fire Protection Association is an international, nonprofit, membership organization dedicated to fire prevention and protection. One of its major missions is to develop consensus codes, standards, and guidelines to protect against fire. Although NFPA codes, standards, and guidelines may be mandatory in some jurisdictions, they are not legal mandates for fire departments nationwide. Despite the NFPA's lack of legal authority, fire departments must be familiar with their publications, as they are nationally recognized and accepted standards. NFPA documents can and have been utilized in civil cases to indicate what a "reasonable person" would do in a given fire situation. Several NFPA standards and guidelines reference rapid intervention.

A number of sections in NFPA 1500, *Fire Department Occupational Safety and Health Program* (last updated in 1997), reference rapid intervention. Section 6-2, "Risk Management During Emergency Operations," states that rapid intervention teams are essentially risk management tools. The majority of the jobs that fire departments actually perform involve risk management in some way, whether fire prevention related, code related, or incident related. When an incident commander (IC) sends a crew to the roof of a building to ventilate, he is managing risk by providing ventilation that may ease the job of personnel in the building. At the same time, the lives of the firefighters operating on the roof are at risk. Anytime firefighters go to a roof, they chance falling off of or through the deck. The longer they remain on the roof and the longer the fire burns below, the greater that risk grows. The IC manages the risk by controlling how long the crew remains on the roof.

The incident commander similarly makes risk management decisions in determining when and how to use rapid intervention teams. A decision to use existing manpower at an incident instead of calling a dedicated RIT, for instance, is risk management. On the other hand, the IC may determine that an RIT is a better risk management mechanism and always call for one.

According to Section 6-2.1.1, "The concept of risk management shall be utilized on the basis of the following principles":

(a) Activities that present a significant risk to the safety of members shall be limited to situations where there is the potential to save endangered lives.

(b) Activities that are routinely employed to protect property shall be recognized as inherent risks to the safety of members, and actions shall be taken to reduce or avoid these risks.

(c) No risk to the safety of members shall be acceptable when there is no possibility to save lives or property.

Rapid intervention team members, and all firefighters for that matter, would do well to remember these principles, both as they apply to rapid intervention operations and even more so as they apply to routine fire-fighting operations. Rapid interventions are high risk operations. Crews enter areas that have already claimed at least one victim. If a rescue effort turns into a recovery effort, the rescue team must significantly reduce their operation's risks. Simply put, if saving lives or property is impossible, risking safety is unacceptable.

Section 6-4 of this standard, "Members Operating at Emergency Incidents," covers several critical points concerning rapid intervention teams:

Section 6-4.3. Members operating in hazardous areas at emergency incidents shall operate in teams of two or more. Team members operating in hazardous areas shall be in communication with each other through visual, audible or physical means or safety guide rope, in order to coordinate their activities. Team members shall be in close proximity to each other to provide assistance in case of emergency.

This section emphasizes the importance of operating in pairs in hazardous areas and remaining in contact with each other at all times, through any of various means. With a buddy system in place, if one firefighter gets into trouble, the buddy will be there to help. Working in teams is just as important for rapid intervention operations. Rapid intervention personnel should work in pairs or larger groups. The teams should remain in contact with each other at all times, either through line of sight, voice, physical touching, or through search ropes or similar connecting devices. Personnel deep in a building filled with smoke and heat must test these contact methods often. It is crucial that buddies do not become separated under any condition.

The NFPA standard continues:

Section 6-4.4. In the initial stages of an incident where only one team is operating in the hazardous area at a working structural fire, a minimum of four individuals is required, consisting of two individuals working as a team in the hazard area and two individuals present outside this hazard area for assistance or rescue at emergency operations where entry into the danger area is required. The standby members shall be responsible for maintaining a constant awareness of the number and identity of members operating in the hazardous area, their location and function, and time of entry. The standby members shall remain in radio, visual, voice, or signal line communications with the team.

Section 6-4.4.1. The "initial stages" of an incident shall encompass the tasks undertaken by the first arriving company with only one team assigned or operating in the hazardous area.

Section 6-4.4.2. One standby member shall be permitted to perform other duties outside of the hazardous area, such as apparatus operator, incident commander, or technician or aide, provided constant communication is maintained between the standby member and the members of the team. The assignment of any personnel, including the incident commander, the safety officer, or operators of fire apparatus, shall not be permitted as standby personnel if by abandoning their critical task(s) to assist, or if necessary, perform rescue, they clearly jeopardize the safety and health of any fire fighter working at the incident. No one shall be permitted to serve as a standby member of the fire-fighting team when the other activities in which he/she is engaged inhibit his/her ability to assist in or perform rescue, if necessary, or are of such importance that they cannot be abandoned without placing other fire fighters in danger.

Section 6-4.4.3....The full protective clothing, protective equipment, and SCBA shall be immediately accessible for use by the outside team if the need for rescue activities inside the hazard area is necessary. The standby members shall don full protective clothing, protective equipment, and SCBA prior to entering the hazard area.

Section 6-4.4.4. When only a single team is operating in the hazardous area in the initial stages of the incident, this standby member shall be permitted to assist, or if necessary perform, rescue for members of his/her team, providing abandoning his/her task does not jeopardize the safety or health of the team. Once a second team is assigned or operating in the hazardous area, the incident shall no longer be considered in the "initial stage," and at least one rapid intervention crew shall be required.

Section 6-4.4.5. Initial attack operations shall be organized to ensure that, if upon arrival at the emergency scene, initial attack personnel find an imminent life-threatening situation where immediate action could prevent the loss of life or serious injury, such action shall be permitted with less than four personnel when conducted in accordance with Section 6-2 of this standard. No exception shall be permitted when there is no possibility to save lives. Any such actions taken in accordance with this section shall be thoroughly investigated by the fire department with a written report submitted to the fire chief.

Firefighters must remember the two-in/two-out rule. Even during initial operations in the hazard zone at a structural fire-fighting scene, they must follow this rule. At the very least, when two responders will be operating in the hazardous area, one of the "out" team must be ready with turnout gear and SCBA to rescue or intervene for the entry team at all times. In deference to departments that are lightly manned, particularly during initial operations, the second out-team member can be the pump operator or incident commander or perform other tasks, as long as he or she clearly accounts for the "in" team at all times. These personnel, however, cannot be standby personnel if abandoning their usual tasks to intervene in a sudden, unexpected event will significantly risk anyone operating at the incident. When the first arriving unit is engaged in structural fire fighting or other hazardous duty, a dedicated RIT must be established quickly once backup units arrive. This will free the initial out team to perform other necessary duties, but a chain of accountability must be maintained so that assigned RIT units can account for the location of all personnel in the hazard zone.

Many firefighters contend that this section of the standard is too stringent to permit discretion when fire crews with limited manpower arrive at incidents where individuals are trapped. Section 6-4.4.5 specifically addresses this risk management issue. Firefighters must quickly analyze the risk and determine if they actually can make a save without placing themselves at an extraordinary risk such as not having another rescue team in place. Only when immediate action can prevent serious injury or loss of life is this acceptable, and in those cases, reporting the circumstances to the department chief in writing is critically important.

Section 6-4.5 deals with emergency medical care for firefighters. It recommends that at least basic life support care and transportation be on standby at a every incident, with advanced life support preferable. If EMS is not routinely available to fire departments at working incidents and the RIT unit is certified and equipped to perform basic or advanced life support, it may fulfill this requirement. At a minimum, RIT personnel should be trained in basic first aid and cardiopulmonary resuscitation (CPR), with higher levels of EMS training preferred. Some form of EMS transportation must also be available.

Section 6-5, "Rapid Intervention for Rescue of Members," stipulates:

Section 6-5.1. The fire department shall provide personnel for the rescue of members operating at emergency incidents if the need arises.

Section 6-5.2. A rapid intervention crew shall consist of at least two members and shall be available for rescue of a member or a team if the need arises [Figure 3.2]. Rapid intervention crews shall be

Figure 3.2 A rapid intervention crew staged and ready for action.

fully equipped with the appropriate protective clothing, protective equipment, SCBA, and any specialized rescue equipment that might be needed given the specifics of the operation under way.

Section 6-5.3. The composition and structure of rapid intervention crews shall be permitted to be flexible based on the type of incident and size and complexity of operations. The incident commander shall evaluate the situation and the risks to operating teams and shall provide one or more rapid intervention crews commensurate with the needs of the situation.

Section 6-5.4. In the early stages of an incident, which includes the deployment of a fire department's initial attack assignment, the rapid intervention crew(s) shall be in compliance with 6-4.4 and 6-4.4.2 and be either one of the following:

> (a) On-scene members designated and dedicated as rapid intervention crew(s)

> (b) On-scene members performing other functions but ready to redeploy to perform rapid intervention crew functions. The assignment of any personnel shall not be permitted as members of the rapid intervention crew if abandoning their critical task(s) to perform rescue clearly jeopardizes the safety and health of any member operating at the incident.

Section 6-5.5. As the incident expands in size or complexity, which includes an incident commander's requests for additional resources beyond a fire department's initial attack assignment, the rapid intervention crews shall upon arrival of these additional resources be either one of the following:

> (a) On-scene members designated and dedicated as rapid intervention crews

> (b) On-scene company or companies located for rapid deployment and dedicated as rapid intervention crews

Section 6-5.6. At least one rapid intervention crew shall be standing by with equipment to provide for the rescue of members that are performing special operations or for members that are in positions that present an immediate danger of injury in the event of equipment failure or collapse.

These sections show the flexibility of the rapid intervention team's makeup and setup and the importance of evaluating risk and establishing an appropriate RIT. (This text covers those issues elsewhere.) The RIT on the scene will need to work with the safety officer and incident commander to continually monitor whether the rapid intervention organization is adequate for the size and complexity of the incident. The RIT unit is in the best position to analyze its ability to respond to likely sudden, unexpected events that might occur at the incident. Moreover, they must ask for assistance whenever additional resources might be needed should something go wrong.

Another standard of interest is NFPA 1561, *Emergency Services Incident Management System,* most recently updated in the year 2000. Section 4-1.9 stipulates, "The emergency services organization shall provide personnel for the rescue of individuals operating at emergency incidents if the need arises. A rapid intervention crew shall consist of at least two individuals and shall be available for rescue of personnel if needed." Elsewhere this standard emphasizes the importance of accountability, personnel tracking, and rehabilitation. Its wording clearly parallels that of NFPA 1500, and the comments on NFPA 1500 also apply here.

OSHA

The Occupational Safety and Health Administration creates and enforces regulations to protect the American workforce. Current OSHA regulations that apply to firefighters include 29 CFR 1910.134, *Respiratory Protection,* which requires employers to provide respirators suitable for the purpose intended and to establish and maintain a respirator protection program. The standard also states that if firefighters must enter an area that is immediately dangerous to life and health (IDLH), at least two must enter together and remain in visual or voice contact with one another at all times, as well as with personnel outside the IDLH atmosphere, such as a working structural fire or hazmat incident. (OSHA defines an IDLH atmosphere as one that poses an immediate threat to life, would cause irreversible adverse health effects, or would impair an individual's ability to escape from a dangerous atmosphere.) The standard also requires personnel who conduct interior fire fighting to use SCBA. In addition, at least two properly equipped and trained fire fighters:

- must be positioned outside the IDLH atmosphere
- must account for the interior team(s)
- must remain capable of rapid rescue of the interior team(s)

OSHA's November 13, 1998, Interpretation to 1910.134(g)(4) to Mark Schultz, senior fire inspector for the Gallatin, Tennessee, Fire Department, further clarifies the requirement for the number of standby personnel:

...the incident commander has the flexibility to determine whether more than two outside firefighters are necessary when more than two firefighters go inside. In a situation where the burning structure is very large, additional outside firefighters may be warranted to ensure effective assistance and rescue. For example, where the fire fighting involves entry from different locations or levels, two outside fire fighters may have to be stationed at each point of entry. You also asked whether standby personnel had to wait for additional standby personnel before entering to attempt a rescue of fire fighters in a structural fire. No. There is an explicit exemption in the Respiratory Protection Standard that if a life is in jeopardy, the two-in/two out requirement is waived. The incident commander and the firefighters at the scene must decide whether the risks posed by entering an interior structural fire prior to the assembly of at least four firefighters is outweighed by the need to rescue victims who are at risk of death or serious physical harm. There is no violation of the standard under rescue circumstances.

The OSHA regulations clearly require a minimum of two personnel to remain at the ready outside whenever firefighters enter an IDLH atmosphere. The standby personnel must maintain accountability for the personnel operating inside. They must be trained to rapidly retrieve responders operating inside, and they must remain capable of doing so, should an emergency develop.

CONCLUSION

OSHA develops and enforces regulations based upon NIOSH research. Depending on the state and the locale, OSHA regulations may or may not apply to a given emergency response organization. OSHA does not consider volunteers as employees and, therefore, does not cover them. But with the variety of compensation arrangements and other municipal ties, OSHA may or may not actually apply to any given fire department. Currently, 24 states, along with the U.S. Virgin Islands, have state plans, in which state agencies administer regulations equivalent to or stricter than OSHA requirements. Again, these plans may or may not apply to particular public agencies. OSHA's December 15, 1998, Interpretation to 1910.134(g)(4) to Senator Jeff Bingaman of New Mexico, clarifies the agency's position with respect to enforcement of this standard among firefighters:

> The "two-in/two-out" policy is part of paragraph (g)(4) of OSHA's revised respiratory protection standard, 29 CFR 1910.134. This

paragraph applies to private sector workers engaged in interior structural fire fighting and to Federal employees covered under Section 19 of the Occupational Safety and Health Act. States that have chosen to operate OSHA-approved occupational safety and health plans are required to extend their jurisdiction to include employees of their state and local governments. These states are required to adopt a standard at least as effective as the Federal standard within six months. The extension of this standard to volunteer fire fighters is a matter decided by each State and is often dependent on whether volunteers are considered "employees" under State law.

NFPA utilizes the consensus process to develop standards and guidelines that help fire protection professionals manage fire problems. Municipalities generally do not adopt NFPA regulations; however, because these standards are all recognized as "good practice" and generally make good sense, fire departments have all the reasons in the world to know and implement them. [2]

Notes

1. The authors in no way represent any of the organizations discussed within this chapter and are in no position to formally interpret their codes or standards. The intent of this chapter is to make emergency responders familiar with the codes and standards that apply to rapid intervention. In turn, responders can decide for themselves how these codes and standards apply to their operations, and they can develop procedures and guidelines so that their actions reflect the spirit of these regulations.

2. Interpretations that apply to the Respiratory ProtectionStandard (and other standards) can be found at the OSHA web site, www.osha-slc.gov/OshDoc/Interp_data. The August 1998 OSHA publication *Questions and Answers on the Respiratory Protection Standard* can be found at www.osha.gov.

Chapter 4
Operational Concepts

Operational Concepts

Rapid intervention concepts are not difficult, and the various codes and standards lay out the general framework for a rapid intervention system. Each department, however, must develop and implement the details of its particular system. Two key decisions are how and when to dispatch the RIT and how to cancel them if they are not needed. The type of vehicle to use to deliver the team's services and how to equip it are also important. Finally, equally important issues are whom to train as RIT members and how to train them so that this service can be provided most effectively.

Dispatch

A significant question for departments to consider is when to dispatch the RIT. In practice, almost every conceivable timing is utilized, from an automatic dispatch on all house and building fires along with other high-risk incidents, to dispatch only at the incident commander's request.

Rapid intervention teams are most likely to arrive on scene quickly when they are dispatched on the initial assignment. In higher risk situations (house or building fires), departments typically assign an additional ladder (or truck) company as the RIT on the initial dispatch. Other departments may charge a heavy rescue company with this function on the initial assignment. To implement the two-in/two-out rule as rapidly as possible, still other departments assign the RIT function to the second arriving engine company on these high-risk calls and assign a third engine company to perform the responsibilities that might previously have been assigned to the second due engine company. While this does quickly provide the RIT, it can also result in many additional "boomerang" calls (smells and bells investigations, smoke scares, and false alarms, for instance), where the RIT unit finds itself being recalled en route to the call or on the scene without ever going into service. This can result in extra wear and tear on the apparatus and personnel, particularly in volunteer organizations. Many departments may not have the resources to commit this extra unit on a routine basis. Others might have the resources but be unwilling to tie them up or risk red lights/siren responses to unnecessary calls.

A logical compromise calls for dispatching an RIT whenever the call is likely to be a working fire or incident. The basis for this decision varies somewhat from department to department. Some consider a call to be a working incident when the dispatch center receives two, three, or more telephone reports of the same situation. Essentially, the additional callers confirm that a serious incident is occurring. Other departments wait for either a police officer or a fire officer on the scene to confirm that the incident is, indeed, working. Some fire chiefs stipulate that only a fire officer actually on scene can confirm the working incident before RIT dispatch. While these practices result in longer dispatch times than sending the RIT on the initial assignment, the delay typically is brief, as major incidents normally result in multiple phone alarms or are clearly significant to responding police and fire officers. On the other hand, this dispatch guideline will minimize unnecessary calls for RIT units.

Yet another dispatch criterion might require an on-scene fire officer's report that "all hands are operating." In this case, the incident commander is indicating that the entire first alarm assignment is committed to a working incident and that no personnel are available on the scene to perform RIT duties. At this point the situation clearly is high risk, with no operating safety plan, and an RIT is needed on the scene as soon as possible. This criterion relies entirely on the incident commander's remembering promptly to send this message, and it requires the communications center to clearly understand the message's importance. The communications center then must dispatch the RIT unit swiftly, and finally the unit must respond quickly to establish an adequate safety net for this incident. An "all hands operating" message may also signal communications personnel to fill empty stations, alert command and/or staff officers, notify utilities, and perform other tasks that might divert their attention from getting the RIT unit on the scene. While this procedure will almost certainly eliminate boomerang calls, it may also delay rapid intervention setup just when it is most needed.

A variation on the all-hands-operating dispatch policy is to automatically dispatch the RIT upon striking a second (or another predetermined number) alarm on multiple alarm incidents. This offers the same advantages and disadvantages of the all-hands policy, with an even greater delay in dispatching and setting up the team. This places first-alarm units at a greater risk if something should go wrong, and it presents the incident commander with the disquieting possibility that incident management might go downhill very quickly should a sudden, unexpected event occur.

Finally, some departments dispatch the RIT only at the request of the incident commander. This relies on the incident commander to see the "big picture," as well as to clearly track resource status and account for all personnel operating on the scene. If the incident commander can do this and has the resources to deal with sudden, unexpected events as they occur, this may be a viable option. Like other options, however, it relies on the incident commander always to remember to request the RIT *before* any sudden, unexpected event occurs. A variation might authorize the communications center to prompt the incident commander to request the RIT if needed on apparently high-risk situations with a majority of the alarm assignment in service. This procedure, however, presents many opportunities for human error that could result in the rapid intervention team's being in the wrong place when it is most sorely needed.

Every emergency response organization needs to determine how the rapid intervention team best fits into its emergency management structure. Decisions must be made based on manpower, resources, and other local conditions that determine the best time to have an RIT on the emergency scene. Once this is established, each organization should establish and enforce strong, clear policies to ensure that the RIT is set up and available before any sudden, unexpected event occurs.

Cancellation

Once the RIT is en route, it may be recalled should the situation come under control. The incident commander must consider, however, that even after an incident is controlled, a number of risks to personnel remain on the scene. A collapse, cardiac arrest, or exposure to fire gases are all quite possible even after incident control. Instead of recalling the RIT, the incident commander may consider having them proceed in at nonemergency speed. In this scenario, the RIT may also be prepared to perform functions such as establishing the rehab sector, lighting the fire ground, performing salvage, filling SCBA bottles, and carrying out other support tasks. This makes the RIT unit quite versatile. The RIT crew may be used to relieve operating personnel or conduct overhaul. Any of these tasks may diminish the RIT crew's ability to perform their designated duties, and the incident commander must carefully consider them during the ongoing size-up and risk analysis. Some departments may designate the RIT crew to set up regardless of the situation and stand by until the emergency is clearly over, only then taking up and returning to quarters. In departments that are already low on manpower, however, this may not be sensible.

Vehicle

Several types of vehicle may be used to provide rapid intervention team services. Some organizations feel that an engine company works best as their rapid intervention team (Figure 4.1). Engine companies are readily available (almost every fire station houses one), and it is simple to add the next due engine company to the assignment as the RIT. Another simple option is to designate one of the engines due on the initial alarm as the RIT team and dispatch another engine to fill the role of the unit now performing RIT duties. Engine companies carry the hose complement that may be needed to lay a backup supply line or to stretch a protection or extinguishment line to make a rescue. Engine company members, however, may not have the highest level of search and rescue training within the organization, and depending on the situation and when they are dispatched, they may not be able to approach the scene closely enough to actually use their hose complement. Also, they may not have the key equipment necessary for RIT duties or the room on their apparatus to store it.

Other departments use a ladder company for their rapid intervention unit (Figure 4.2). They simply add an extra ladder company to perform

Figure 4.1 Engine companies are readily available for dispatch for RIT duties.

Figure 4.2 Ladder companies generally have the training and equipment to perform rapid intervention.

these duties at building or working fires. Ladder companies are generally well equipped and trained to perform search and rescue duties. They bring a solid complement of devices to rescue victims from heights, and they typically carry saws and other heavier extrication tools. A disadvantage of this concept, however, is that ladder companies may not be readily available, particularly in more rural areas. In many areas of the United States, ladder company service is 30 minutes or more away. In these areas, a ladder company realistically cannot perform rapid intervention team activities in any reasonable time frame. Even where ladder companies are available, they may be limited and have to provide more traditional tactical services on the fire ground, as opposed to being held in reserve for RIT duties.

A number of large and medium-sized fire departments have formed heavy rescue companies to rescue trapped civilians and firefighters in serious incidents (Figure 4.3). There are normally fewer rescue companies than engines and ladders, and they are brought to the scene to handle the toughest jobs. They typically bring properly trained and equipped personnel to handle even unusual rapid intervention team assignments. Though they might not carry ladders and hose lines, rescue personnel usually have experience with engine and ladder companies and can gather needed equipment from those units already on the scene. Rescue companies are generally even less readily available than ladder companies, however, and in other areas may be trained only to perform medical, vehicle, or other specialized rescue functions and lack the fire-fighting experience necessary to handle rapid intervention duties at fires.

Other options for providing rapid intervention are varied. In some volunteer or paid-on-call departments, members respond to the scene in their personal vehicles. In this case, a rapid intervention crew could be formed from members gathering at the scene. The necessary rapid intervention equipment must be carried either in the members' personal vehicles or, perhaps, in a specialized compartment on a response vehicle. Some departments use a "flying squad," which is a personnel unit that responds to working incidents across the jurisdiction (Figure 4.4). Flying squads may

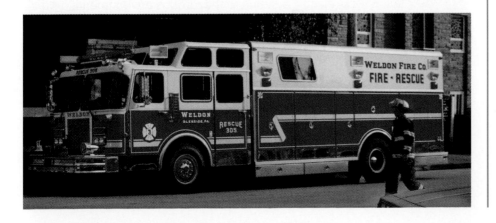

Figure 4.3 Many rescue companies were created to serve the original need for rapid intervention.

Figure 4.4 The flying squad provides the manpower cushion for rapid intervention.

use small, quick vehicles that cannot carry the rapid intervention equipment but can quickly bring personnel to the scene. Optimally, the RIT vehicle should be able to carry at least a minimal assortment of rescue gear. Lighting equipment with a power generation unit is also quite practical. A hybrid ladder/rescue unit may be the optimal platform from which to deliver rapid intervention, but these may be available only to larger, well-equipped departments. Some version of a rescue company, such as a special service or light-to-medium rescue unit that does not carry heavy rescue equipment, may effectively provide rapid intervention in all but the most complicated rescues.

Each department must evaluate its apparatus and personnel and dispatch the right equipment to perform rapid intervention team services to every specific incident.

Sample RIT Vehicle Equipment

A vehicle routinely used for rapid intervention team services might carry the following primary equipment (Figures 4.5 and 4.6):

- Command post equipment and workstation with cell phone hookup

- Thermal imaging camera with wireless remote video and AC/DC TV/VCR fixed in vehicle or mobile

- Hazmat library

- Ventilation fans (electric suitable for hazardous environments, and 24-inch gasoline powered PPV)

- 20 kW PTO generator with two 250-foot cord reels

- 20-foot 6000 watt Will-Burt light tower

- Two 1500 watt tripod lights

- Four 500 watt portable lights

- Hazmat operational level equipment (diking materials, decon equipment, Class D extinguisher, hazmat identification kit)

- Basic life support responder equipment/automated defibrillator

- Triage/mass casualty tarps and tags
- Stokes basket
- Rehab equipment, water jugs, towels
- 4-bottle 6000 psi cascade system
- Stihl circular rescue saw, primary setup with metal cutting blade
- LEL/O_2 and CO meters
- Ventilator/heater with 25-foot duct for venting confined spaces and providing warm air for entrapped victims
- Radiation emergency equipment

The variety of equipment makes this particular vehicle quite versatile when responding to working incidents. It provides equipment to support EMS, rescue, fire, and hazmat operations. The vehicle can support rehab, air supply, and other operations, particularly once the incident is under control. Crews equipped with light-to-intermediate-level rescue equipment must also be resourceful in obtaining more advanced equipment to handle more complicated situations. They must not only know where to obtain these resources, but must also train with them so they can be integrated seamlessly when necessary. Departments with greater resources might carry the more advanced equipment on a heavy rescue or ladder company that has the majority of tools immediately at hand.

Figure 4.5 A rapid intervention vehicle with equipment complement.

Figure 4.6 A light tower is an excellent resource for an RIT unit.

Equipment

RIT equipment can be divided into two main categories: personal equipment carried by all RIT members, and team resources that will be staged or collected on scene. Departments already own most of this equipment, both personal and collective, which will not cost additional money. The RIT company can often acquire additional needed resources on scene from other units. Companies may need to look for outside sources to fund more expensive devices such as thermal imaging cameras. Federal, state, and municipal government representatives typically are open to requests for life-saving RIT equipment that not only benefits the local community but also protects the lives of rescuers from several jurisdictions. In pursuing such funding, keep in mind that firefighter rescue equipment can also serve the civilian population in many other applications.

Personal Gear

The rapid intervention crew's basic equipment is similar to that used for structural fire fighting. Full turnout gear, including a protective hood and gloves, is a must. Because team members may need to perform rope rescue and work with small accessories, they should also carry leather rescue gloves in an easily accessible pocket.

All members of the team must wear SCBA the entire time that units are operating on the fire ground; however, they do not need to be "on air" until actually deployed. If available, SCBA should have extended-use air tanks that provide at least 30–45 minutes of air in case difficult extrication situations arise (Figure 4.7). Larger departments in which all firefighters wear the same model air packs may opt for a buddy-breathing system, although sharing air with victims will quickly reduce the rescuers' operating time under hazardous conditions. Smaller departments providing RIT for mutual aid companies will likely find many shared air attachments to be incompatible with one another and might decide to carry spare SCBA or supplied air respirators instead of buying buddy-breathing devices.

Other equipment carried by individuals will depend on department resources and personal preference. If possible, all team members should have a portable radio. Each person's ability to monitor and send transmissions increases the likelihood of hearing a call for help, as well as the team's versatility. While not every member needs a light, each group should carry portable hand lights to facilitate interior and night operations.

To further increase the team's ability to operate safely and successfully during a rescue, all members also should

Figure 4.7 Extended-use SCBAs provide for greater RIT versatility.

carry at least some personal survival gear. Although departments will have different standard operating guidelines regarding this equipment, firefighters will benefit from having door chocks, medical trauma shears, personal rope bags or rescue slings/webbing, and a life-safety belt or harness. (Some manufacturers now incorporate rescue harnesses into their turnout gear and jumpsuits.) Team members can use this equipment not only for their personal security and self-rescue, but also for the rescue of downed firefighters.

Team Equipment

Although rapid intervention teams may depend on other apparatus for larger and specialized resources, they should carry essential equipment on the team vehicle. One of the most important rapid intervention items is rope. It is used not only for lowering downed firefighters to safety, but also as a search line for guiding rescue crews.

Figure 4.8 Rope bags allow for the easy storage and deployment of lines.

While there is no set number of ropes an RIT should carry, each team should have at least two—one for search and one for rescue. Although having just two ropes keeps things simple, additional lines will increase the team's readiness and versatility by allowing prerigging for various scenarios. Regardless of the number of ropes on the apparatus, all should be stored in rope bags. These nylon or canvas carrying sacks often come with draw ties and backpack-style shoulder straps that allow rescuers to carry multiple ropes without tangling them (Figure 4.8). Rope deployed from a bag comes out smoothly and evenly, which is a real advantage if it must be dropped to the ground from a ladder or out a window. Once the team has practiced and prepared for a number of rescue evolutions, individual rope bags can be prerigged and labelled for rescues requiring specific knots. Some teams color code their rope bags to distinguish between life-safety and utility rope and then sew labels onto each bag for easy identification. The most accessible and clearly marked rope must be the search rope.

When creating a search rope, individual departments have many options, depending on local preference. The most easily acquired search line is a regular utility rope; however, most teams choose a stronger, thicker rope with several modifications. All RIT ropes should have knots at regular intervals to indicate the distance traveled. The distance between knots can vary, but it is usually between 10 and 20 feet (Figure 4.9). Each end of the rope should be tied with a loop and have an attached carabiner. This will allow team members to secure the rope to fixed objects both inside and outside the building. If necessary the carabiners can be removed and used during extrication. Some departments may further modify the search rope

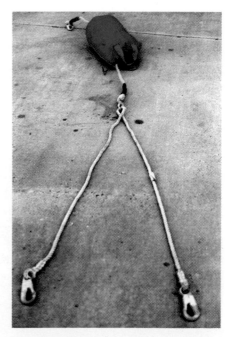

Figure 4.9 This search line allows RIT members to move up to 20 feet apart from each other and into rooms.

by placing a ring on its end and connecting two 10-foot ropes (Figure 4.10). The added ropes can enable the team to cover a larger search area or allow one firefighter to enter and search a room while the other stays at the door. This setup permits team members to move up to 20 feet apart while working from the same search rope. If the department does not use split rope, team members can carry 10-foot personal ropes that can be clipped onto the main line.

Since RITs depend so heavily on their ropes, they must maintain and inspect them regularly in accordance with NFPA 1983, *Standard on Fire Service Life Safety Rope and Components*. For further discussion of rope types, care, and inspection see the current edition of *Essentials of Firefighting* (Stillwater, Oklahoma: IFSTA). When repacking or preparing RIT ropes in bags, the team should take its time to ensure that they carefully coil the rope in the bag for quick deployment.

Aside from ropes and personal gear, other essential RIT equipment includes a variety of forcible entry, cutting, and striking tools (Figure 4.11). At a minimum these will comprise a set of irons (axe and Halligan bar) with which team members can improvise numerous fire ground functions. If available, a large selection of hand tools should be staged with the team. These could include Kelly bars for more forcible entry power, a variety of hooks, a battering ram, a pry bar, and bolt/wire cutters.

RITs should also consider power tools. These can greatly reduce the team's physical labor, leaving them fresher for victim removal. Saws are the most readily available power tools for RITs. If possible, teams should carry or have available to them both a chain saw and a circular saw with multiple blades. If operating at a house fire, the chain saw will likely be most useful. If necessary, the circular saw can be fitted with a wood blade. Together these saws will be available for tasks such as breaching walls or removing debris. Teams operating at commercial or industrial occupancies will probably rely more on circular saws fitted with blades for cutting metal or concrete. If the RIT vehicle has only one circular saw, it should be stored with the metal cutting blade in place. Together with the chain saw, that will give the team ready saws for occupancies constructed of either wood or metal (Figure 4.12).

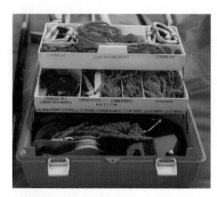

Figure 4.10 Some RIT units carry a box of carabiners, slings, and other accessories to increase rescue flexibility.

Figure 4.11 RIT members need a variety of tools in order to perform efficiently when called to action.

Figure 4.12 The proper saw can hasten access to trapped firefighters.

In addition to saws, many other power tools may be of use to the RIT. One example is the hydraulic Rabbit Tool, which is similar to a Porta-Power and is one of the most effective forcible entry tools available (Figure 4.13). This compact device exerts about 3–5 tons of force at its spreader tip and can open almost any door in minutes. It is especially useful in occupancies with multiple locked doors, such as apartment complexes or office buildings. Other useful power tools may include air bags for lifting debris, an air chisel, and a cutting torch.

Numerous other pieces of equipment may be desirable. The team is limited only by the resources available on the fire ground or in the region (extra equipment can always be dispatched). From other units at the scene, the RIT should be able to access a variety of ladders, hoses, vent fans, spare SCBA, lighting equipment, and more specialized rescue gear.

Teams must choose their own equipment for each rescue, given the occupancy involved and the severity of the fire. Regardless of what they bring to the incident, they must keep it in a designated RIT staging area, preferably on an appropriately marked tarp *(RIT* or *FAST,* for example). When laying out the equipment, the team should coordinate it by type, so rescuers can easily find specific tools. For example, ropes should be together, with any labels facing up. Saws should be beside each other with their blades laid out. To ease carrying tools, the team can also stage two firefighter gear bags. The bags serve as excellent carrying cases for small equipment such as cutting shears (which can be taped to the bag), blankets used for dragging or covering victims, a rescue SCBA, multiple ropes, hand lights, and power tools that need to stay clean (Figure 4.14).

Training

RIT members' training varies greatly. For example, if the region's dispatch protocol requires the second engine to be the RIT, the team's qualifications will depend on whoever happens to be riding that vehicle. On the other

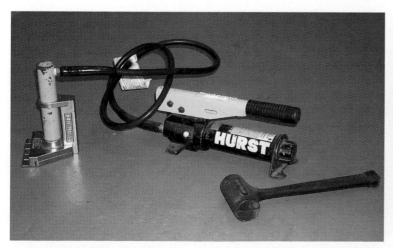

Figure 4.13 The Rabbit Tool is a compact and powerful entry device.

Figure 4.14 Firefighter gear bags allow for easy organization and transportation of small and awkward RIT equipment.

hand, where protocol designates certain companies or departments as battalion or regional teams, they can establish more specific training requirements. In setting qualification standards for team members, chief officers should consider their area's occupancies, local resources, and the number of personnel who could or would complete the training.

Individual Training Requirements

At a minimum, RIT members should meet Firefighter I requirements (NFPA 1001). This will qualify them to perform basic search and rescue, forcible entry, and ladder operations, as well as to use ropes and SCBA. Beyond these fundamentals, courses such as building construction, basic rope rescue, collapse rescue operations, structural fire rescue, firefighter survival, vehicle rescue, emergency medical technician or first responder, and Firefighter II (NFPA 1001) will strengthen team members' abilities. Those seeking even further enrichment should consider classes such as incident safety officer, advanced rope rescue, and advanced collapse rescue operations. Ideally, all regional fire academies should offer both awareness and operations level RIT courses to teach firefighters not only how to rescue one another but to greatly improve the safety of the fire ground as a whole (Figure 4.15).

Department Training Requirements

In addition to courses offered at the state/provincial and regional levels, station drills on RIT are vital. If companies do not actively practice their skills, they may not be able to perform them under the stressful and difficult conditions of a real rescue. Using resources such as this book, company and department training officers can ensure that rapid inter-

Figure 4.15 Training facilities must incorporate RIT programs in their curriculum.

Figure 4.16 Regular company drills must incorporate RIT concepts and skills.

vention procedures and techniques are regularly reviewed under realistic conditions (Figure 4.16).

Companies should make every effort to practice RIT operations using live "victims" in real SCBA. (If the equipment officer complains, drills might use an older pack that can be treated roughly.) Rescuers' should always wear all equipment, including gloves, and their masks should be covered or blacked out. To add realism, members watching the drill can provide background noise, such as running PPV fans, to simulate fire ground sounds.

RIT training requires physical facilities much like those already found at many fire training institutes. Ideally they include search and rescue buildings with smoke or live fire, mazes, interior and exterior stairwells, roof or floor trapdoors, a variety of windows of various heights, and areas for wall-breaching practice and forcible entry simulation. If nearby facilities do not exist, departments can construct or improvise many rescue situations on their own.[1] They might use empty basements in local buildings or houses scheduled for demolition, or they might build a basic drill tower and maze. Existing facilities within stations are also potential training grounds. Team members can practice many basic search and rescue techniques in an empty engine bay, and most stations have stairs where members can practice dragging victims up and down.

Conclusion

Dispatch and response policies for RITs vary from jurisdiction to jurisdiction, as do the vehicles that deliver RIT service. Training may also vary, depending upon available instruction at emergency services academies. The basic equipment and drills that RIT personnel should receive, however, are generally quite standard. Much of this equipment and training should already be available to the average firefighter. Forcible entry and search and rescue equipment and training should be a staple of every fire department. Where it is, slight training modifications can ensure that firefighters are used to search for their downed comrades as well as for civilians. Those departments where it is not a staple must focus on these areas.

Note

1. On building an inexpensive RIT training facility, see Rick Lasky, "Saving Our Own: Designing A Firefighter Survival Training Aid," *Fire Engineering*, May 1998. This article includes an overview, plans, pictures, and usage suggestions. For an example of a regional approach to RIT training, see Michael Robertson, "Safety in Numbers," *Fire Chief*, May 2000.

Chapter 5
On-Scene Organization and Preparation

On-Scene Organization and Preparation

Being ready to go to work is a critical rapid intervention function. This includes being prepared for prearrival tasks, arrival and setup, scene size-up, and communications. The incident commander must also understand his role and the RIT's place in the incident command system when a firefighter is downed or missing. The team must also be prepared for rapid intervention at special operations.

Pre-Arrival Tasks

From dispatch, the RIT members must prepare for deployment. En route, the officer should perform or delegate several tasks related to initial size-up and preparation. Many firefighters become trapped during the early phases of a fire, and the RIT may be required to deploy soon after arrival.

As with any other unit, the RIT's incident size-up should begin at dispatch. The officer should consult any dispatch notes to determine the type of fire and the time of the initial alarm. If a preincident plan for the fire building is available, the officer also should note the building's construction and any specific dangers and consider how the burn time will affect structural collapse and the volatility of on-site hazards.

To obtain further information about the scene, monitor radio transmissions carefully. Be sure to identify incident operating channels and to determine which units are on location, where they are assigned, the extent and intensity of the fire, and any indications of potential hazards. If more than one operating channel is being used at the incident, the officer may wish to assign individual members of the team to monitor each channel separately. Additionally, he should note nonoperating channels that are available for local talk-around use. Firefighters may use these channels to call for help from fellow company members without alerting incident command. This may avoid the perceived embarrassment of calling attention to their need for help.

Arrival and Set-up

Upon arrival, the RIT team must move efficiently to ensure maximum readiness in a minimum of time. The officer should immediately go to the command post and determine the officer to whom the team should report. That individual should then advise the team where to stage and what radio frequencies to use. The RIT staging area is typically near the command post, although at larger fires and high-rise operations it may be elsewhere. Once the team has this information, they can begin to set up an equipment area. While the officer meets with the IC regarding size-up, the team should prepare for deployment (Figure 5.1).

Figure 5.1 Initial contact with the IC must be made upon arrival.

On-Scene Organization and Preparation

To clearly identify their staging area and prevent the use of their equipment by other firefighters, the RIT unit should deploy a large tarp (Figure 5.2). They can then lay out their tools according to use and need. While members prepare the staging area, the unit's driver can survey the scene for larger resources such as aerial apparatus placement, adequate water supply for hose line protection, and a variety of ladders.

Once staging is ready, at least two team members should maintain a high level of readiness. They should have their tools either on their person or close at hand, their SCBA bottles on, and their PASS devices activated. Depending on local procedure and the type of SCBA used, RIT members may want to have their face pieces and protective hoods in place to further reduce deployment time.

Scene Size-up

To facilitate the size-up and preparation process, companies performing rapid intervention functions should devise a written checklist. This permits the team to prepare specifically for the hazards at the particular incident. The RIT officer may break the checklist into parts, delegating various size-up duties to team members (Figure 5.3). The checklist may also include staging hazard-specific tools for the situation at hand or performing additional duties according to local protocol. While the RIT officer reports to the command post, the team prepares their equipment and continues their size-up. The RIT officer at the command post and the team sizing up the hazards may use separate checklists or information gathering tools (for sample lists, see appendices 1–4).

At the command post, the RIT officer should confirm prearrival information on burning time, the building's construction and age, and other pertinent details. He also should ascertain the location of all crews,

Figure 5.2 The RIT equipment area is denoted by a large tarp.

Figure 5.3 Team members must perform an ongoing and detailed building size-up.

their assignments, the extent of the fire, and on-site hazards. If the rapid intervention team does not already have a preincident plan of the building, the command post might have one. The officer should also determine whether a personnel accountability system is in place and locate any tactical worksheets related to the incident.

While the team leader reports to the command post, the team should begin their survey of the building or incident site. They should circle the fire building, completing a checklist to note both traditional elements of size-up and those specifically related to RIT operations. These should include:

- Occupancy size and possible rescue concerns

- Structural instability and collapse hazards

- Fire progress, especially through areas of truss construction

- Access points to each area of the building, including the basement and roof

- Access obstacles

Obstacles such as barred windows and locked doors pose extreme risk to firefighters and should be noted. The RIT team also must assure that the resources to gain access at these points are assembled.

As the survey continues, the team should note firefighter locations. Members should observe and report freelancing to the IC, watch windows in case trapped firefighters appear, and note and immediately report any uncontrolled utilities. They should also assess ground conditions and watch for signs of firefighter entrapment such as helmets, pillows, or other out-of-the-ordinary debris on the ground around the building. They also should determine if the terrain or the weather poses any specific challenges to potential rescues. For example, uneven ground will make ground ladder operations difficult, while snow may make roof conditions especially dangerous.

Once team members complete their initial building survey, they should share all results with the rest of the team. The team should then make plans regarding access points and specific rescue scenarios. As the incident continues, so should size-up. The officer or a senior team member should regularly survey the building to assess conditions and report any differences in fire behavior, low-air alarms, PASS device activations, changes affecting access, or any other significant developments. Proper and continuing size-up not only improves the RIT team's likelihood of success in case of rescue, but may actually prevent the need for rescue in the first place.

Communications

Communications are vital to the RIT's effectiveness. The team must be able to clearly monitor ongoing fire ground transmissions and quickly detect distress calls. This becomes difficult if multiple departments are operating, particularly when they are on different channels or, sometimes, even on different radio bands. The RIT officer must detail one team member to monitor radios, which may be this individual's sole assignment until the team springs into action. If multiple channels are being used, the officer may assign a second team member to monitor auxiliary frequencies.

The RIT must also maintain effective internal communications. They need to be able to deploy swiftly and efficiently, and this is best coordinated by good communications within the team. Solid team communications will also enhance the team's safety when members are deployed into a risky situation. To facilitate this, the RIT team generally should operate on a frequency other than the fire ground channel in use at the incident. This allows team members to communicate across the fire ground without interfering with other transmissions. Depending on local preference and the number of channels/radios available, the RIT can develop any number of communication strategies. One option is to pair team members, assigning one member of each pair to monitor a fire ground channel while the other sets his radio to an RIT operating frequency. The firefighter with the RIT radio would communicate directly with the RIT officer, who would in turn communicate directly with his superior at the command post.

In the event of a firefighter distress call, command should immediately clear all traffic except the IC, the RIT, and perhaps, the victim, off the operating band. If the downed firefighter has radio capabilities, he may be able to transmit his location. Even if he is lost, notification that a rescue plan is underway may reassure a distressed firefighter and calm him down. During a rescue, the RIT officer must stay in direct contact with command. If the team officer must enter the building, a liaison, such as the driver, should stay outside to monitor and relay communications.

Incident Commander's Duties

A missing or downed firefighter is among the most difficult, stressful situations that any incident commander can face. In many departments, the individual in trouble may be a relative or a close friend. The IC's first priority must be to maintain self-control. Otherwise, he has no way to control other responders operating at the scene.

The IC needs to gather as much information as possible about any missing or downed firefighter, so that proper actions can be taken to locate and rescue the individual. He must obtain this information extremely rapidly from the personnel who were most recently with the person who is in trouble. This size-up will allow the proper action plan to be implemented. Who, what, and where are all key questions at this point. Also, the primary hazard should be quickly identified: is it fire, collapse, or some other danger? What resources are needed to continue operations safely?

What initial actions can be taken without exposing anyone else to unnecessary danger?

Once the IC answers these questions, he must reevaluate priorities and assign tasks. Should the incident go defensive? Will the building have to be sacrificed? Should ventilation, fire attack, forcible entry, laddering, or other tactics be modified to protect and support the rescue efforts? Do additional sector officers need to be designated to ensure safety? What resources will be needed to effect the rescue? Is rescue likely, or is the operation potentially a recovery? Tasks must be assigned, including committing the RIT. EMS must be alerted to prepared for one or more patients. If tasks are not assigned, others will assign themselves to the tasks. While this is going on, a personnel accountability report (PAR) should be initiated. It will clearly identify who is accounted for and who is not.

The IC should seriously consider another alarm. Even if the additional units are not needed immediately, an extended rescue operation can quickly tax resources. If a staging area has not already been established, he should stage these resources nearby. During the downed firefighter size-up, the IC must clearly understand that the stakes have been raised significantly. Others in the hazard area are at increased risk. Withdrawing some or all other companies operating at the scene may be prudent, but this decision may be difficult to enforce, particularly with a brother down in the hazard zone. A standardized rescue plan that the department practices and uses regularly will ease this somewhat. Everyone's highest priority should be not to lose anyone else.

Fit into ICS

Each department needs to determine where the RIT best fits within its incident management system. Many departments find that the RIT operates best under a safety officer, reporting to him directly rather than through the incident commander (Figure 5.4). Even the RIT's basic functions are closely aligned with the responsibilities of the safety officer. They have more freedom to operate anywhere on the fire ground and more authority to rectify any safety problems they might identify. In some cases, the RIT officer may be designated as the scene safety officer if the RIT is appropriately manned and its operations will not be affected adversely. One disadvantage of the RIT's reporting to the safety officer is that information may not flow smoothly if the safety officer is highly mobile at the scene. This could also affect team accountability.

RIT functions may also fall under the operations officer (Figure 5.5). In particular, at hazmat incidents, the hazmat team will likely develop its own RIT, which will report directly to the hazmat sector officer, who in turn reports to operations. In emergencies where the RIT is assigned to the operations area, it makes most sense for them to report directly to the

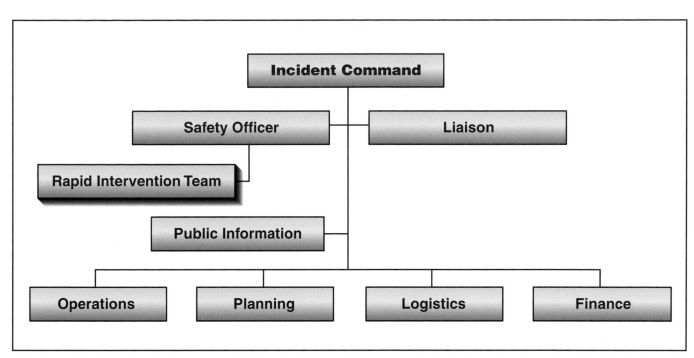

Figure 5.4 RIT Reporting to Safety Officer.

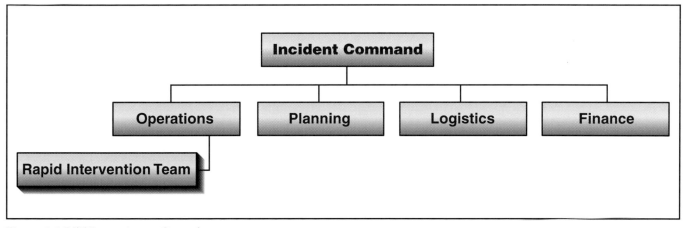

Figure 5.5 RIT Reporting to Operations.

operations officer. This will allow the most rapid deployment of the team and the best management of the situation should a sudden, unexpected event occur. One potential disadvantage of this arrangement is that the RIT could be drawn into the incident if the operations officer does not maintain strong control over resources assigned to that branch. Some departments may feel more comfortable assigning the RIT to the staging officer (Figure 5.6). This clearly keeps the RIT apart from operations, but like any other unit staged, allows them to be pulled quickly into the situation. Because staging typically implies that the unit is not immediately at the emergency scene, though, this could potentially delay the actual deployment of RIT personnel in a real event.

At lesser incidents, none of the officers in these examples may be present. It is then most logical for the RIT to report directly to the IC (Figure 5.7). The IC may prefer to have the RIT report directly to him for

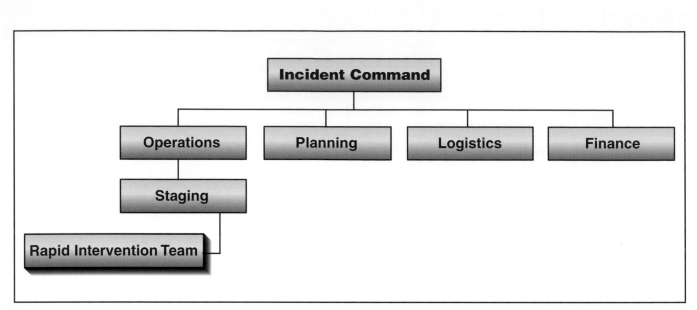

Figure 5.6 RIT Reporting to Staging.

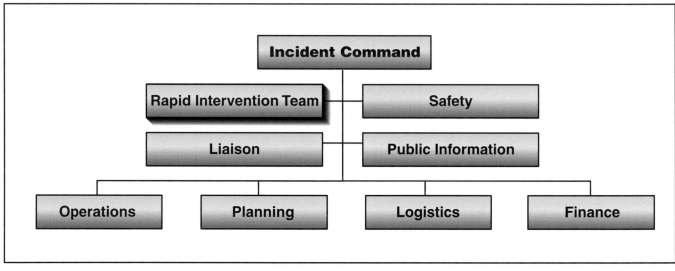

Figure 5.7 RIT Reporting to Command

other reasons, as well. Information flow and accountability for the scene will be clearly communicated to the team, and the IC can deploy them without delay once he is aware of a firefighter in trouble. On more complex incidents, however, maintaining a team liaison at a well-developed command post may be difficult. The IC must consider all of these possibilities when determining where the RIT fits into any particular incident's management system.

Special Operations

Rapid intervention operations require that personnel be fit mentally, as well as physically, to meet the challenge. Making top-notch firefighters stand by at a working incident is difficult. Yet their function at that scene is no less critical to the operation's success than that of the firefighter on the hydrant.

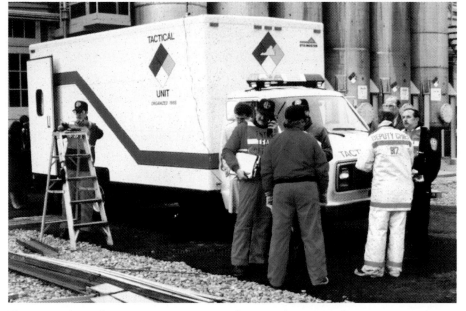

Figure 5.8 Special operations require special approaches to rapid intervention.

Firefighters who will be assigned to rapid intervention crews must understand this up front. Like airport crash-fire-rescue personnel, they do an awful lot of standing around and waiting. They stand by and watch operations occur, never knowing when things will go wrong and they will have to react immediately to the situation to save the lives of others. If that is not stressful enough, when the RIT does go into action, it is likely because another firefighter or emergency responder is in trouble. Crews assigned to rapid intervention have a good chance of coming across a firefighter who has been severely injured or killed on the job. They must be prepared to deal objectively with this high-stress situation and perform at their peak to extricate the victim from his predicament. For this reason, some fire departments prefer to have mutual aid personnel perform rapid intervention duties, as they are somewhat less emotionally tied to the situation and can evaluate it more objectively. At the same time, officers in charge of rapid intervention crews should not hesitate to take advantage of critical incident stress teams and resources, which their personnel will very probably need in order to maintain their mental fitness.

Hazardous materials teams generally implement a kind of RIT. When two hazmat team members suit up for entry into the hot zone, two members also will be suited up in the cold zone, prepared to rescue the personnel in the hot zone if needed. In some cases, the standby personnel will not have staged the equipment needed to remove the entry team, nor will they have practiced actually removing personnel wearing entry suits and SCBA. This is a much different ballgame than rescuing civilians or even firefighters in turnout gear, and it is certainly a procedure that hazmat RITs should practice. Fire-fighting rapid intervention teams who are not properly trained or qualified must make this clear if summoned to a situation that requires hazmat duty.

Confined space operations are another area where policy already dictates the rapid intervention team's organization. Confined space protocols typically call for one rescuer to stay outside of the confined space at all times. The rescuer outside must monitor hazards and be able to retrieve a downed entrant with a hoist or similar arrangement and to call for additional rescue assistance as needed. Obviously, during confined space rescue operations, a rescue team should be in place and ready to extricate rescuers who have entered the hazard area; however, firefighters trained in RIT operations may not be the best individuals to perform this type of standby. Confined space rescue standby personnel must be well trained in extricating victims from these special hazard areas. Trenches, too, can be considered confined spaces, although ventilation may be a bit better than in an enclosure. Trench rescue emergencies may also call for a specialized rapid intervention approach.

Collapse rescue is another complex emergency. In major events, urban search and rescue (USAR) teams are likely to be mobilized. These units have RIT teams built into their operations. Smaller collapses might require a reduced response, but an RIT operation should be in place for these events as well. Personnel trained in fire-fighting rapid intervention also should have the basic skills to provide rapid intervention during these lesser collapses.

Wildland/brush fire incidents require a different approach to rapid intervention. Unless timber collapse is likely, RIT operations will involve either trying to reinforce or to quickly rescue units threatened with being overrun by fire. These incidents usually stretch resources, and any spare resources may not be able to mobilize quickly enough to pluck potential victims from a fast-moving wildland fire. Nonetheless, ICs at these incidents must consider the potential for a sudden, unexpected event and keep some resources in reserve for rapid deployment to mitigate the situation or at least to render first aid and rescue to units that may fall victim to the incident.

Conclusion

Quite conceivably, different units can be assigned to RIT operations, depending on the hazard involved. Fire-fighting RIT operations may be assigned to one company or group of companies, while hazmat, confined space, and other specialized operations require specialized RIT personnel to stand by.

Performing rapid intervention team duties is not as simple as it may seem. The team must deal with every call as if it were a drill and with every drill as if it were a call. The team must continually practice simulated evolutions to ensure that operations will go right the first time when they are called to action. Being ready requires a number of tasks, each as important as any other. The team must understand their place in the scene's incident management system and then fit into their role like a hand into a glove. At any point during the incident, something could go wrong. It is for these moments that the RIT must constantly maintain their peak readiness. Preplanning and advance practice alone will prepare the team to meet this difficult challenge.

Chapter 6
Deployment: The Two-Team Rapid-Intervention Concept

Deployment: The Two-Team Rapid-Intervention Concept

Chapter 6

Even with the best equipment and size-up, rapid intervention teams cannot fulfill their primary purpose of rescue unless they are mentally and physically prepared to deploy immediately upon receiving a call for help. The teams should be ready for a wide variety of entrapment situations and, therefore, must maintain a high level of versatility in order to adapt to rapidly changing fire ground conditions. Most importantly, they must be prepared for the possibility that they themselves might become trapped.

Given these requirements, a *minimum* five-person, two-team rapid intervention unit is ideal. Dividing the unit into at least two teams essentially delegates setup, size-up, and rescue tasks to specific team members. This frees the officer from having to micromanage and empowers the firefighters to take a proactive, involved approach to tasks at hand. Furthermore, providing at least two-teams ensures an RIT for the RIT, remaining true to the two-in/two-out concept.

Two-Team Overview

The two-team RIT action plan's primary components are a lightly equipped, agile entry team and a multipurpose support team. Together with an outside safety and communications person (either the officer or driver), these teams can be put to use in a variety of ways, depending on fire ground conditions and circumstances. Upon arrival, the two firefighters designated as the entry team will lay down a tarp at the RIT staging site. They must work quickly to prepare their tools in case they must perform a rescue in the early stages of the incident. Preferably kept on the unit's truck, first team tools typically consist of the firefighters' personal equipment plus the search rope, a set of irons (preferably on a strap to allow hands-free travel), hand lights, and a thermal imaging camera if available. Once set up, the first team must maintain a high level of readiness. At this point, they can conduct the

Table 6.1
RIT Assignments

Primary Team	Secondary Team	Outside Person/Team
• Assemble first team equipment	• Assemble all other equipment at staging area	• Determine on-scene resources
• Maintain a high state of alert	• Perform building survey and safety functions	• Can perform size-up
• Enter structure following Mayday call	• Provide additional manpower for entry team	• Maintain search rope at entrance
• Perform interior victim assessment	• Bring additional equipment to entry team	• Communications during rescue
• Perform/assist with rescue	• Assist entry team from exterior	• Gather/provide additional manpower if needed

building survey to familiarize themselves with the layout and report hazards to the officer (Table 6.1). Any other tasks should be limited to those that do not require strenuous physical exertion, such as monitoring radios.

While the first team prepares, the second team gathers the remainder of equipment to be staged on the tarp and then sizes up ladder and hose resources. Once the second team completes these tasks the officer can deploy them to deal with scene safety tasks and ongoing size-up around the building. If the officer is a part of the second team, the unit's driver can be assigned temporarily to work with the second team's firefighter to carry out initial tasks while the officer reports to command.

In the event of a rescue, the RIT officer will rapidly brief the entry team, and they will move in with the search rope to assess the emergency and determine whether additional resources will be necessary. If the entry team finds that they can easily perform the rescue, they may do so (for example, firefighters requesting help because they are lost and need to be led out). The second team remains in place while the fifth RIT member monitors the search rope from the entry point. If, upon reaching the victim, the entry team finds a situation requiring rescue operations, they can radio to the outside personnel for additional resources. Equipment can be brought quickly into the scene following the first team's rope. The second team may also operate from the exterior to remove victims or bring materials. The fifth member remains at the entrance to the hazard area and maintains communications. In the event of second team deployment, the RIT officer should consider requesting additional personnel to perform exterior tasks such as monitoring the search rope or placing ground ladders.

Having a second team outside the building allows the first team to enter rapidly, with a minimum of equipment. Following a report from the first team, the second team can select tools from the large assortment of staged equipment to resolve most any potential fire ground rescue situation. If the second team is not needed to perform the RIT operation, it serves a vital use as the first team's RIT while ensuring a safer exterior environment.

Mayday

Knowing when help is needed is of fundamental importance to the RIT operation's success. Even with teams constantly monitoring fire ground frequencies, calls for assistance can be easily missed or not understood. Departments and mutual aid companies need to jointly determine radio and scene protocols regarding how firefighters in distress should request assistance and what actions to take immediately upon receiving such a call.

All firefighters should know exactly how to communicate a need for emergency assistance. Some agencies use the word *Mayday* to indicate a firefighter in distress. In the event that a Mayday call is transmitted, it should be repeated on all operating channels. This will alert others to the gravity of the situation and might allow existing resources to rectify the crisis immediately, prior to deployment of the RIT. Command must immediately request radio silence and have the RIT prepare to deploy. The IC should also try to determine the incident's exact location, the number of victims, and their predicament. If radio contact with the victims is not possible, command can call for any personnel who witnessed the incident. A personnel accountability report (PAR) must then be conducted to confirm interior reports and to ensure that no other firefighters are lost or trapped. As soon as the IC can determine the location or general area of the problem, the RIT is deployed. Further information can then be relayed to the team as it is obtained.

While the RIT deployment and accountability checks take place, it is vital that other teams working on the fire ground continue their assigned tasks — especially in the early stages of an incident, when manpower is limited. If a hose team leaves its position, more firefighters could become endangered. Additionally, an abandoned primary search could result in civilian deaths. In many cases, discipline among those working on the fire ground during a RIT deployment will result in a better outcome for all. Achieving this demands regularly incorporating Mayday procedures into training exercises and making personnel accountability reports a part of routine incidents.

In addition to the individual Mayday call, the RIT can use other methods to deploy at an incident. The incident commander may notice something unusual or a particularly hazardous situation and be unable to contact personnel operating in that area. He may call upon the RIT to enter this situation and extract the personnel at risk. While the personnel may not be victims yet, this assignment will still place RIT members in danger. Precautions must be taken to minimize the risk. The RIT may also be deployed when a PAR indicates a responder is missing. This is a time when

> **MAYDAY**
>
> NFPA 1561, *Emergency Services Incident Management System,* 2000, discusses the Mayday, or emergency traffic, concept:
>
> > Section 2-2.4.1. In ensuring that clear text is used for an emergency condition at an incident, the ESO (Emergency Services Organization) shall have an SOP that uses the radio term **emergency traffic** as a designation to clear radio traffic. Emergency traffic can be declared by an Incident Commander (IC), Tactical Level Management Component (TLMC—commonly known as division, group or sector), or member who is in trouble or subjected to an emergency condition.
> >
> > Section 2-2.4.2. When a member has declared an emergency traffic message, that person shall use clear text to identify the type of emergency, change in conditions, or tactical operations. The member who has declared the emergency traffic message shall conclude it by transmitting the statement **All clear resume radio traffic**.
> >
> > Section A-2.2.4.1. Examples of emergency conditions could be "Fire fighter missing," "Fire fighter down," "Officer needs assistance," "Evacuate the Building/area," "Wind shift from north to south," "Change from offensive to defensive operations," or "Fire fighter trapped on the first floor." In addition to the emergency traffic message, the Emergency Services Organization can use additional signals such as an air horn signal for members to evacuate as part of their SOPs.
>
> Each department must develop standard operating procedures (SOPs) on calling for help in an emergency. This can be as simple as the procedure outlined in the NFPA standard. Once the SOP is developed, it is critical that each member clearly understand it. All personnel must be trained in the procedure and practice it again and again so that they will know how to call for help when they are in trouble and know how to react if someone else calls emergency traffic. Routine departmental training sessions should incorporate emergency traffic drills so personnel can actually act out and critique the procedure. Having a common, clear term to call for help will assist the RIT with identifying situations that call for their deployment.

the RIT needs to rely on their size-up and communications with the IC/sector officer to pin down the missing firefighter's possible location. Of course, the team should deploy quickly when an actual entrapment is witnessed, but they must always ensure that the IC is notified of all RIT actions.

Entry Team Operations

Once command deploys the RIT, the entry team must move as quickly as possible to locate the victim. Because they use a search rope and typically have some idea of where they are headed, the entry team does not have to move in as strict a pattern as a traditional primary or secondary search team. With the aid of a thermal imaging camera, the team can move even more directly.

The entry team should begin by securing their search rope outside the immediate hazard zone (Figure 6.1). For a house fire, this will likely be at an exterior door. For high-rise or commercial fires, anchor points may be found in stairwells or at fire doors. The fifth team member assists by securing the rope and flaking it out, extending it with additional ropes if necessary, by checking to ensure entering teams are wearing all gear properly, and by pulling the rope tight if needed for removal. The rescuer at the door also serves the vital function of team accountability.

After entering, the first team's members need to work together as much as possible. The more they have practiced as a unit, the smoother they will operate. Depending on the design of their search rope, the rescuers may wish to spread out with part of the rope between them in large areas. For small room searches one team member can wait at the door, flaking the rope while the other sweeps the room at the rope's end (Figure 6.2). At each room or area entered, the team should stop and perform a visual and audio survey. While one member scans the area with the thermal imaging camera, the other can listen for firefighter calls or PASS devices. To increase the likelihood of hearing or seeing a victim, the team may stop completely and hold their breath.

Figure 6.2 Often the most efficient search technique is for one member to perform a small room search while his partner monitors the egress point.

Figure 6.1 Search lines should be secured at the entrance to the hot zone.

While advancing, the entry team must be sure not to move so rapidly that they miss areas where victims might be concealed. Firefighters buried in rubble often will not appear on thermal imaging cameras, and in such instances only a hands-on search will suffice. When the team reaches the target area, they must take special care both to find the victim and to ensure their own safety. The hazard that entrapped, injured, or disoriented the victim will likely still be present when the RIT arrives.

USING THERMAL IMAGING

Thermal imaging cameras (TICs) have rapidly changed the nature of search and rescue in the fire service (Figure 6.3). Firefighters should, however, understand that TICs have limitations and cannot replace a solid foundation in basic structural search techniques. While in many cases TICs have greatly reduced the time required to find civilian victims, in some instances their use has delayed locating victims because the search team did not follow a basic pattern or did not fully understand the camera's operation. Whenever you use a TIC, remember the following points:

- TICs' view screens offer an image of an area's heat differentiations; they do not truly "see in the dark." In some cases a TIC's screen may appear blank if the team finds itself in a room of heavy smoke that is all roughly the same temperature. This will limit the RIT's ability to navigate if they are not already following a search pattern of some sort.

- TICs cannot "see through" most objects. RIT rescuers using a TIC might see a buried firefighter only if his body is sticking out from the debris. The team should probe debris so as not to miss a victim.

- TICs can reflect thermal patterns off mirrors or glass, causing rescuers to appear as other firefighters. Energy-efficient windows can also reflect interior heat patterns, thus concealing possible exit/rescue points.

Figure 6.3 Thermal imaging cameras provide a strong advantage for rapid intervention work.

- A TIC's screen shows heat patterns relative to the overall amount of thermal (heat) energy in the room. Areas that look very hot on the imager may not really be very hot at all; they are merely hotter than other items in the room. (Newer models can give specific temperature readings.) Team members should not base a hazard analysis solely on the color of their imager's screen.

Thermal imaging cameras can serve a vital use to RIT teams by identifying potential hazards before the team reaches them and by making it far easier to locate victims than with conventional techniques alone. Ideally every RIT team, everywhere, will have thermal imaging technology available to them. RIT members must remember, however, that the technology that makes the cameras most useful—their ability to see thermal energy—also gives them certain limitations. For this reason, drills should first be conducted without the use of TICs. The cameras can then be added to the training program in general and to RIT in particular.

Deployment: The Two-Team Rapid-Intervention Concept

Locating A Downed Firefighter

When attempting to locate a downed firefighter, RITs can use a number of techniques to narrow the search. One of the most important of these techniques, ongoing size-up, ideally begins long before a Mayday is called. RIT members can monitor radios and watch fire ground activities to track where firefighters are working within the structure. From observations at the RIT staging area and during building walk-arounds, the team can note the areas where firefighters could most likely become trapped. Note how observing scene operations and monitoring radio orders helps the RIT in the following hypothetical situation:

The RIT arrives on scene to find a dilapidated two-story residential occupancy with smoke and fire showing from the second floor. The incident commander, having been told that homeless people may be inside the occupancy, has begun an offensive attack. A member of the entry team notes a team of firefighters advancing a charged, 2-inch line in the front door and up the main stairwell. Another team places a ground ladder and ascends to the roof while a third team receives radio orders to begin a primary search of the second floor, entering via ladder through an upper window. Suddenly a Mayday call is heard. A member of the primary search team has collapsed and is too heavy for his partner to pull him out. Knowing exactly where the firefighters entered and their orders, the RIT entry team is able to quickly locate them by climbing the search team's ladder and conducting a right-handed search with their rope. Meanwhile the hose and roof teams stay in place to protect the downed firefighter. The next arriving units continue the primary search.

In reality, finding a downed firefighter will not likely be so simple. RIT members may be faced with zero visibility, high noise levels, and intense heat.

Even if exterior size-up provides the RIT with some knowledge of the victim's last location, a variety of interior search techniques may be required. Consider the following methods to further narrow the search:

- *Hose Lines or Ropes.* If the victim is a member of a hose team, try to determine which hose he was operating (this can sometimes be determined by diameter or color-coding). Follow the hose to the firefighter's last location. If the victim is assigned to a search team that was using a rope, simply follow the rope into the building. Some jurisdictions are now color-coding or marking search ropes and hoses to allow easy identification of teams.

- *Radio Contact.* If the fire ground operating channel is clear of traffic, direct contact should be attempted with the victim. He may be able to guide the RIT to his position. If the firefighter is unconscious, radio contact may still be possible. Place two radios together and transmit to create a loud feedback noise. If the victim is nearby, the RIT may hear his radio.

- *Witness Accounts.* Either outside or inside the building, the victim's team members will likely be able to provide valuable information about his whereabouts.

- *Listen Carefully.* Like civilian victims, trapped or disoriented firefighters may call out for help. The entry team can pause occasionally and call to the victim, waiting silently for a response. Other noises that may indicate the victim's location include an active PASS alarm, regulator/breathing sounds, tapping noises, moaning, and shouting.

- *Watch Carefully.* Firefighters requiring RIT assistance may leave visual clues. Pay attention for discarded gear, tools, or ropes. In lighter smoke conditions, a victim's flashlight beam may be visible.

- *Remember Firefighter Survival Training.* When searching, especially for a disoriented firefighter, try to think as the victim might. Consider the various firefighter survival techniques taught in your region. For example, many firefighters are taught to move to a wall or window if disoriented. Knowing this, RITs can move rapidly along walls to find a missing firefighter.

Even using the above techniques, finding trapped firefighters may be very difficult, especially if they are unconscious. The RIT needs to remain as calm as possible to avoid overlooking the victim in haste. Teams must practice over and over until they can be certain that they are operating at maximum efficiency, with the proper balance of speed and thoroughness. At the same time they must work to become skilled at mentally mapping an occupancy as they search, in order to call additional resources if necessary. When they locate the victim, seconds may count.

Victim Assessment

Once the entry team reaches the downed firefighter, they must perform a rapid patient assessment, similar to an EMS initial trauma survey (Figure 6.4). Like an arriving medical unit, the RIT must ensure scene safety to avoid becoming victims themselves. They should be especially concerned with the presence of entanglement hazards, live wires, secondary collapse, and advancing fire. Only after establishing its own security should the team turn its attention to the victim.

Like EMS, RIT teams should initially perform a modified ABCD survey (airway, breathing, circulation, deformity) when they reach a victim. Begin as with CPR, assessing consciousness verbally and by a sternal rub if necessary.

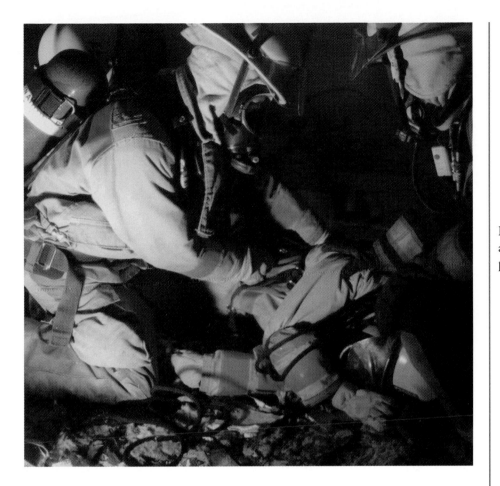

Figure 6.4 The initial RIT victim assessment resembles a typical EMS primary survey.

- *Conscious Firefighters.* Assure them by stating you are there to help and that they must stay calm and not move. If possible, determine the level of air in their SCBA cylinders. Be sure to assure them that should their low-air alarm sound, the RIT is prepared to provide them with air via buddy-breather or spare SCBA.

- *Unconscious Firefighters.* Place your ear over their SCBA regulators and determine if they have an air supply. Note the quality of breathing – is air free-flowing through the regulator or are they respiring regularly? If possible check the level of air in their tanks to determine if rescue will be feasible before their air runs out. (Do not rely on a low air alarm; some may not sound because of very low pressures or because their whistle/bell is damaged). If the patient is not breathing, the team must perform an immediate removal.

Securing the Firefighter's Air Supply

Without an air supply in a toxic environment, a victim's survival chances are remote. Checking a downed firefighter's SCBA is thus one of the most vital tasks during the initial victim size-up. If the victim has no SCBA or if the air in his SCBA cylinder is dangerously low, the RIT must immediately decide on action that could mean the difference between life and death for the victim.

If the initial size-up indicates an air supply emergency, the first team has two main options. They can decide to quickly extricate the downed firefighter themselves or stay where they are and try to secure an air supply. If they choose rapid extrication, they must quickly access a means of egress, such as a window or door. They will not have time to cross obstructions such as stairs that must be climbed or debris that may entangle them. If immediate egress is not possible, the team must consider several air supply solutions:

- *Buddy Breathing.* Probably the quickest method to get air to the victim, buddy breathing requires that the victim and the RIT have either the same type of SCBA or an adapter to make them compatible. One rescuer simply takes the buddy-breathing hose from his own air pack and hooks it into the downed firefighter's SCBA. The victim then breathes from the rescuer's air supply. Obviously, this operation will quickly deplete the rescuer's SCBA, especially if the victim is breathing heavily.

- *Spare SCBA.* A simple method of bringing air to the victim, a spare SCBA provides a fresh separate air supply for the downed firefighter. The entry team either takes in the new air pack during the initial RIT search or calls for it when the size-up indicates its necessity. The second team can either drag the unit by its straps or place it in a gear bag to avoid entangling the harness or damaging the regulator (Figure 6.5). The easiest way to transfer air to the victim is to remove the victim's regulator and plug in the fresh SCBA's regulator in its place, then secure the pack to the victim for removal. To ensure compatibility of air packs, the RIT should try to determine what type of pack the victim is wearing before they enter the structure with a replacement. If the regulators are incompatible, the team will need to replace the entire mask. They must be sure the new pack is charged and ready to go in service. Finally, when they reach the victim they must perform the exchange as quickly as possible using two team members, one to remove the old mask and another to place the new one (Figure 6.6).

If securing the victim's air supply with a fresh SCBA, the team must secure the second SCBA to the victim before moving him. The easiest way to do this is to place the fresh SCBA on top of the victim's body and then feed its waist belt through the victim's SCBA shoulder straps. If necessary, the RIT may cut the empty SCBA off entirely and replace it with the new one. They can do this by rolling the victim on one side, securing one shoulder strap, and then rolling him to the other side to secure the other strap. This technique is time consuming and should be used only if absolutely necessary. If the victim is only a short distance from the exit, the team may simply have one rescuer carry the pack while the others move the victim.

Figure 6.5 Some RITs carry a spare SCBA unit, including adapters and facepiece.

Figure 6.6 Two team members work together to provide a rapid air supply switch.

- *Supplied Air Respirator.* Although somewhat specialized, the supplied air respirator (SAR) is the best method for sustaining a long-term air supply to a downed firefighter. Supplied by large air tanks on a truck outside the building, the SAR consists of an air hose that the team takes to the victim. They connect it either directly into one of their own regulators or to a separate mask that they place on the victim. Once hooked to an SAR, a downed firefighter caught in an advanced entrapment scenario (such as a roof collapse) can breathe fresh air for hours until freed. The SAR's major drawbacks are its set-up time and potential damage to the air supply hose along its length through the fire building. Preferably, the assembly of any SAR used will include one or two back-up 5-minute air bottles in case the hose becomes compromised.

Patient Information and Other Considerations

Presuming the downed firefighter is breathing, conscious, and has an air supply, the team should obtain the following information:

1. The victim's name and company and whether the victim is alone

2. If the victim is injured or having a medical problem

3. The nature of the victim's collapse, entrapment, or disorientation

This should take no more than 15–30 seconds. Once the RIT has determined the number of victims and the general situation (or if the victim is unconscious), the RIT should conduct a head-to-toe preextrication survey. Since RIT operations often take place in noisy and dark environments, the entry team must be sure to quickly run their hands over the victim to determine the following:

- Is the firefighter entangled in any way (such as cable wrapped around the SCBA)?

- Are there signs of traumatic injury such as a fall? Consider the firefighter's body position. Does it look as if he slumped over or landed with impact?

- Are any objects on or under the victim? For instance, is his leg caught under a heavy timber or is any sharp debris under his body?

- Does the firefighter's gear reveal obvious signs of fractures, lacerations, or punctures?

- Does the firefighter complain of any pain or injury as you proceed?

Determining Action

As soon as the team completes the rapid survey, it must consider how to remove the firefighter. If a traumatic injury is indicated, the team must contemplate using spinal immobilization, such as a backboard. If interior conditions are good or improving, bringing medical personnel and resources to the victim before extrication may be advisable. In the event of a deteriorating hazard situation, RIT should use whatever method of removal will be least harmful to the firefighter. In essence, the RIT must perform a rapid risk analysis to weigh the possibility of further injury from movement against the chance of injury from interior conditions. Generally, if the downed firefighter is in cardiac or respiratory arrest, the analysis is moot. The team must move the firefighter as quickly as possible to a location where CPR and defibrillation can be applied safely.

Before putting their plan into action, the first team must report the following information to the team officer:

- The situation—the number and condition of firefighters who need help

- Any entrapment issues

- Whether the team will need additional resources to complete the rescue

- What the team's next action will be

A sample report could be, "Entry team to RIT, we have one firefighter down. Exit will be by window on side C. Prepare for ladder removal."

In some cases the entry team may decide to remove or treat the victim on their own. This is especially true if the rescue is related to a medical collapse or if the firefighter is disorientated rather than entrapped. Before acting without additional support, the entry team should ensure they can safely and quickly move the victim the needed distance. If they must traverse stairs or large areas, the entry team should consider backup as a safety precaution.

The Second Team

The second team is the RIT's insurance policy. The "cavalry," this team stays outside until the entry team specifically requests them. Because they remain outside, these two firefighters are able to bring almost any resources needed to the entry team, either via the interior or exterior. In large structures or if the entry team cannot find the victim, the second team can deploy to search other areas of the building. Should the entry team themselves become trapped, the second team acts as their RIT.

Although the second team may perform fire ground safety tasks as the entry team stands by outside the structure, they should move to the RIT staging area as soon as a Mayday is received. As the entry team goes in, the second team should check their own personal protective equipment to ensure they are ready for immediate deployment if necessary. Importantly, they must monitor radio traffic and continue to size up conditions the entire time that the entry team is in the structure.

If the entry team requests any tools, the second team assembles them for rapid movement. If the entry team only needs help with lifting or dragging, the second team may bring slings or a blanket for a drag. Prepacked rope and equipment bags are highly recommended. A sample second team rescue bag might include a life rope with prerigged pulley, a strobe or other light to mark exits or the victim, trauma shears, a blanket for dragging, and slings. An assortment of small tools could also be carried in this bag. With these tools, the second team rapidly follows the entry team's rope to the victim.

If called upon to support exterior operations, the second team may need to move a significant amount of equipment to the egress point. They should do this as efficiently as possible. For example, if carrying a ladder, the team may safely and easily carry ropes and other equipment in bags with shoulder straps. Before leaving the RIT staging area, the second team should make sure it has all necessary equipment. The entry team inside may not appreciate wasted trips between the rescue site and the staging area.

MI: The Number One Fire Ground Killer

Although it is vital that RITs train for difficult firefighter extrication scenarios, the most likely use of RIT is for a firefighter suffering a heart attack. Given that heart attacks cause up to one half of all firefighter deaths on the fire ground, RITs must be prepared to rapidly remove and treat a firefighter suffering from heart attack or cardiac arrest.

While rescuing a firefighter suffering from a heart attack might not seem as glorious as using air bags to extricate a victim from under a heap of rubble, its urgency can be far higher. If a trapped firefighter is in stable condition and protective hose lines are in place, RIT personnel often can take their time to ensure that they perform the rescue as safely as possible. In the case of a witnessed cardiac arrest on the fire ground, every second counts. As taught in almost all CPR classes, the human brain begins to die in just a few minutes if it does not receive oxygen. Lifesaving cardiac defibrillation must be given within those few minutes to increase the victim's chance of survival.

With the rapidly expanding popularity and availability of automatic external defibrillators (AEDs), one likely will be on the RIT's unit or at least at the scene. If possible, the RIT should stage this vital piece of equipment with the rest of their gear. When a Mayday is received, the entry team must work quickly to determine if the victim is experiencing cardiac symptoms. If he is, the team should take the following steps:

- Reassure the conscious victim and try to keep him calm.

- Remove the victim as soon as possible.

- If the entry team cannot remove the victim and no significant fire-related hazards exist, consider having the second team bring the AED and CPR equipment.

- Have EMS stage as close as possible to the planned exit so advanced life support can begin as egress is made.

- If the victim is in cardiac arrest and EMS is not readily available, the RIT should begin CPR and AED as mandated by their medical training level and local protocols.

Figure 6.7 Rapid intervention personnel must be prepared to perform basic EMS functions.

Clearly, the best way to deal with the number-one fire ground killer is to promote healthy practices among firefighters. Physical fitness facilities and programs can help address the problem before it begins. On scene, performing rehab on all firefighters as they finish their first air cylinder can greatly reduce their physical stress and, therefore, lower the probability of cardiac collapse. If a cardiac emergency occurs, however, RITs *must* be prepared to handle it (Figure 6.7).

Transfer of Firefighter to EMS

Once any rescued firefighter is brought out of the fire building, care is transferred to EMS personnel. Since NFPA standards require an ambulance to respond on all structural fire assignments, EMS should already be on scene. The RIT officer should be sure that this is so and determine their location through the incident commander. In the event of a rescue, the team leader should arrange for EMS to be ready to receive the victim. If ambulance personnel are not available, as when they must transport multiple victims from the scene, the RIT should provide the highest level of medical care they can until the patient's transfer is possible. As soon as medical personnel arrive, care should be transferred, but RIT members can offer assistance as needed.

Conclusion

An RIT action plan will greatly enhance RIT operations on the fire ground. Through riding assignments and prearranged setup tasks, the RIT officer will be able to focus directly on the particular circumstances at hand. A multi-team approach allows for the deployment of secondary assistance and resources from either the exterior or the interior of the occupancy.

In reality, smaller departments may not be able to meet the ideal minimum of five firefighters working in teams. Many will be lucky to have even two personnel available for the RIT. Each jurisdiction must formulate the plan that best suits its own needs. If only two firefighters are on the initial RIT, the team can be supplemented as more personnel arrive. Through the adoption of similar regional standard operating procedures (SOPs) and combined training efforts, RITs can be constructed at the scene if necessary. Remember that RIT teams do not have to be heavy rescue teams trained to the USAR level. They must, however, be resourceful and do whatever is necessary and prudent to save firefighters in trouble. Knowledge of basic rescue techniques and training in particular entrapment scenarios will greatly increase the average firefighter's ability to serve as an RIT member.

Chapter 7
Interior Rescue Evolutions

Interior Rescue Evolutions

Even during the most basic training evolutions, removing a firefighter from a building is no easy task. The added stress and anxiety of an actual downed firefighter situation, combined with the noise, poor visibility, and confusion of a working fire, create a situation that will test the RIT's experience and training to its fullest. Each minute that passes without the victim's being located or freed will add to the tension. Aside from remaining calm, RIT members must be trained to the point where they carry a mental "toolbox" of solutions to common rescue and removal challenges. Many of the solutions found in the next two chapters have been created by fire departments and individuals across North America in response to firefighter entrapment situations.[1]

Ranging from drags and carries to the removal of a firefighter from below grade, this chapter explains techniques of removing the downed firefighter from a building under a wide variety of fire conditions and hazards. They are listed roughly in order from the most common and basic to the more complicated and advanced. The basic skills prerequisite to performing the evolutions are described first, followed by more detailed firefighter rescue techniques.

Rope

Ropes are generally described in terms of the working end, the running end, and the standing part (Figure 7.1). The working end is whichever end of the rope that is used for tying knots. The running end is the opposite end of the rope and generally is used for hauling. The standing part is the remainder of the rope. For tying RIT-related knots, only two terms are necessary—*bight* and *loop*. To form a bight, grasp the rope and bend it back on itself while keeping the sides parallel. To make a loop, first make a bight and then turn the rope so the side of the bight crosses the standing part of the rope (Figures 7.2, 7.3).

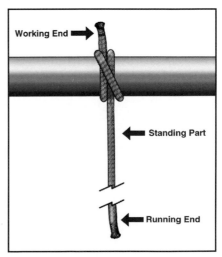

Figure 7.1 Parts of a rope.

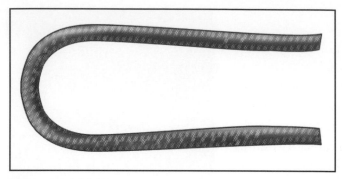

Figure 7.2 A bight.

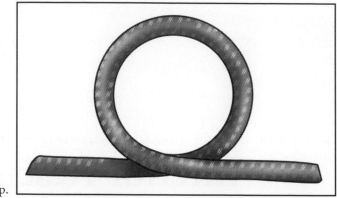

Figure 7.3 A loop.

Figure 7.4 Carabiners eliminate the need for many fireground knots.

RIT members can perform almost all rescues with a minimum of technical rope knowledge and using one main knot, the figure-eight on a bight. This knot is used to create a loop anywhere in a rope (Skill Sheet 7.1). RIT members also may use the handcuff knot, a modified clove hitch, which is useful for securing a rope to a victim's arms or legs in preparation for lifting or dragging (Skill Sheet 7.2). A third, and final, knot used in the practical evolutions is the water knot. This knot is used to tie personal webbing or a sling into a loop (Skill Sheet 7.3).

Once RIT members learn the above knots, they are ready to work with carabiners. These clips are used extensively in RIT evolutions because they are simple to apply, intensely strong, and easy to undo. They also can be securely locked. These features give them a distinct advantage over many knots that are more difficult to use in the dark. Every RIT team should carry at least two carabiners for each of its ropes. If each rope is stored with a figure-eight on a bight at each end, a carabiner can be clipped to each loop (Figure 7.4).

Freeing Entangled Victims

When performing their initial victim assessment, RITs may find firefighters entangled in cable or wire. This is especially likely in buildings that have been wired for computer networks and have large amounts of cable running above the ceiling. To free entangled victims, the RIT can carry trauma shears that quickly cut through most cable. For heavier wires the team can carry wire cutters. If the victim is caught in a large mass of cable, try pulling him feet-first from the entanglement and then cutting the wires.

If a member of the RIT becomes entangled in wires, he should back out of the area. If he must proceed forward, or if backing out does not work, the firefighter should turn to the side to reduce his profile. Large sweeping movements of the arms, as if doing a swimming backstroke, often will prevent a firefighter from becoming entangled. If necessary, advancing rescuers can cut cable as they go or use the sweeping technique to collect them in piles with their arms.

> ## WARNING
>
> Never cut wires unless the power to the building has been secured. Keep in mind that even when the power has been cut to a building, secondary electrical sources, such as generators, may still be operating. If victims are found entangled in wires, they must be approached with considerable caution. Determine if they are conscious before touching them – if the victim is unconscious it could mean an electrocution hazard exists.

Moving the Downed Firefighter

The following techniques are designed for moving downed firefighters through hazardous interior conditions. RITs can face a wide variety of situations that prevent standing, including high heat, confined spaces, and piles of debris. It is vital, therefore, that they know a number of different lifts, drags, and carries. These techniques will provide a foundation for an RIT member's "toolbox" of victim moving skills. For any given rescue, the team might use several of these techniques as they adapt to changing interior conditions. In any event, they should always use the fastest and most efficient technique possible under the circumstances.

Converting the SCBA to a Rescue Harness

To perform any RIT moving technique efficiently and without compromising the victim's air supply, you may first need to loosen the downed firefighter's SCBA straps enough to allow placing ropes or webbing through

Figure 7.5 An RIT member creates a rescue harness using the victim's SCBA straps.

them. Loosen the straps only as much as necessary, and retighten them after you have placed the rope or webbing. This may reduce airway obstruction caused by the victim's body hunching over, and it also can give you more control when lifting an unconscious firefighter's limp body.

Even with the SCBA straps tightened, the pack may ride up the victim's body during the rescue. To prevent this, you can convert the SCBA assembly into a rescue harness. First, unbuckle the waist belt and loosen each half. Next, slide one half of the belt under the victim and between the victim's legs, secure the fastener, and then tighten the straps appropriately. In addition to keeping the SCBA from sliding up the victim's body, this improvised rescue harness will help keep his air supply intact (Figure 7.5).[2]

Using SCBA Straps as Handles

One of the easiest ways to move a downed firefighter is to use his air-pack shoulder straps as handles. If interior conditions permit, one rescuer stands behind the victim, grasps the SCBA straps, and pulls backwards toward the exit (Figure 7.6). The second team member stands at the first rescuer's back, facing the exit, and guides them out using the search rope. As with any lifting, this firefighter drag is best done using the knees and legs to lift and push while keeping the back straight.

If the space around the victim is large enough, two members of the RIT can perform this drag. One rescuer stands facing *toward* the exit on either side of the downed firefighter. Each grasps an SCBA strap, and together they pull the victim (Figure 7.7).

Webbing/Sling-Assisted Drags

This drag is almost identical to the SCBA harness drag, but uses a sling such as those used on circular cutting saws or a one-inch webbing to create

Figure 7.6 A simple SCBA harness drag.

Figure 7.7 The two-person SCBA harness drag can be performed from a standing or crawling position.

leverage. In the basic version of this drag, the rescuer slides his personal webbing under the downed firefighter's SCBA harness, stands up, and drags the victim out, using each side of the webbing as a handle (Figure 7.8). If two rescuers can fit in the space, each can grab either end of the webbing. Some firefighters prefer to keep a loop tied in each end of their webbing. Once they have passed the webbing through the victim's SCBA harness, they slide the loop on the running end of the webbing through the loop at the working end and pull, creating a sealed loop around the SCBA straps. They can then use the loop on the running end as a handle to pull the firefighter.

Figure 7.8 Personal webbing can give a single rescuer greater victim leverage.

Figure 7.9 Creating a dragging harness from a saw sling.

Figure 7.10 This harness provides rescuers leverage when dragging in a forward crawling position.

Figure 7.11 Formation of a girth hitch for dragging.

If the downed firefighter is not wearing an SCBA, the rescuer should consider using a power saw sling or a pretied webbing loop. To create a hauling handle, place the looped sling or webbing over and up one arm at a time, as if placing a jacket on the victim (Figure 7.9). Both sides of the loop pass over the victim's back. To create the handle, slide the bottom strap of the loop under the top strap and pull it through. This technique is useful when dragging a victim without an SCBA through environments that require crawling, such as confined spaces or rooms with advanced thermal layering (Figure 7.10). It is preferable to other non-SCBA sling or webbing drags that place the loop around the victim's chest and, thus, can cause a breathing restriction.

Rope/Webbing Drag

A variation of the sling drags, the rope/webbing drag gives RIT members a line for hauling the downed firefighter while remaining in the crawling position. This method of moving a victim requires two rescuers and a 20–25-foot length of 1-inch webbing or rope. First, the rescuers tie the rope or webbing in a loop. Next, they slide the working end of the loop under the victim's SCBA shoulder straps and then pass the working end through the loop of remaining webbing to form a girth hitch (Figure 7.11). They then pull on the working end until the loop tightens around the shoulder straps. This forms a large loop of webbing that one rescuer can put over his shoulder and crawl ahead of the victim, pulling him with the webbing, while the other rescuer pushes (Figure 7.12).

If the rescue area is small, two RIT members can move the victim in stages. They advance the webbing ahead of the victim, stop, and pull the victim to them. They can then use another drag to round a corner and continue on with the webbing drag. If the RIT wants to pull out the

Figue 7.12 Webbing provides rescuers leverage when dragging in a forward crawling position.

firefighter feet first, they can modify this drag by tying a handcuff knot in the rope or webbing. They then attach the knot to the firefighter's ankles or lower legs and drag the victim as above.[3]

Push-Pull Drag

The push-pull drag is remarkably useful when operating under high heat or in compact spaces. After the entry team performs initial assessment and determines to move the patient on their own, one rescuer moves to the victim's feet and the other rescuer moves to the victim's head. Assuring the victim is on his or her side, the first rescuer places one of the victim's legs over his shoulder and secures it with his arm. Meanwhile, the rescuer at the head grasps the victim's top SCBA shoulder strap. When ready, the rescuer at the head calls "move" and simultaneously pulls while the other rescuer pushes. With the leverage achieved by the rescuer at the feet, this drag will provide rapid movement from the crawling position (Figure 7.13).[4]

Blanket Drag

Often overlooked in the fire service today, the blanket drag remains a highly efficient way to move both firefighter and civilian victims. Familiar to most and included in IFTSA's *Essentials of Firefighting*

Figure 7.13 The push-pull drag provides excellent leverage for rapid removal while crawling.

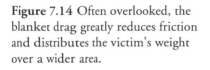
Figure 7.14 Often overlooked, the blanket drag greatly reduces friction and distributes the victim's weight over a wider area.

manual, this drag requires only a blanket. The rescuers roll the victim toward themselves onto his side, gather the blanket in behind the victim along his body, and then roll him back away from themselves onto the blanket. The blanket under the victim lessens friction and provides additional handholds for the rescuers (Figure 7.14).

Fireman's Carry

Another basic fire-fighting lift, the fireman's carry remains useful for firefighter rescues. Used primarily in areas with limited hazards, it allows one rescuer, with assistance, easily and rapidly to remove a victim from the building over long distances. Its main disadvantage is the strength, time, and exertion required to raise the victim over the rescuer's shoulders. To use this lift, one rescuer should stand on or over the victim's legs to keep them from sliding and then grasp the victim's SCBA straps and pull him to the sitting position. The second rescuer stands behind the victim and helps lift

Figure 7.15 The fireman's carry is difficult to initiate, but effective once begun.

him onto the first rescuer's shoulder (Figures 7.15, 7.16). Alternately, the first rescuer can pull the victim to his feet while the second rescuer moves under the victim's abdomen. The second rescuer then lifts the victim onto his shoulders. If RIT team members are to use this lift, they should practice it regularly. Each team member must know how large a firefighter he can lift safely.

Creating a Simple 2:1 Pulley System

As many firefighters are aware, a simple pulley will give the rescuer a two to one mechanical advantage. In other words, when an RIT uses a 2:1 pulley system, dragging the victim takes only half as much power as a regular drag would require. This is especially useful if the team has limited manpower or if the victim is relatively heavy.

To prepare a makeshift 2:1 simple pulley system for dragging, the RIT should follow these steps:

1. If the firefighter is not wearing a personal harness, or if his SCBA frame does not have a hook, place a large carabiner or hook through the SCBA frame or straps (Figure 7.17).

2. Pass the working end of a hauling rope through the hook or carabiner.

3. Secure one end of the rope to a solid object near the point where the firefighter is to be dragged; if necessary this could be a team member.

4. Move ahead of the downed firefighter and pull on the free end of the rope that is not attached to the solid object (Figure 7.18). Additional RIT members can help pull the rope, drag the victim by the straps, or assist with a modified push-pull drag.

Figure 7.16 The second rescuer lifts the victim onto the shoulders of the other team member.

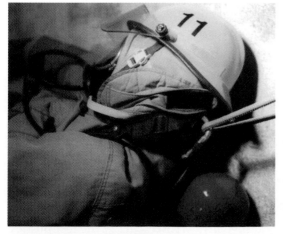

Figure 7.17 Use a carabiner to connect the hauling rope to the victim's SCBA.

Figure 7.18 The pulley drag provides excellent leverage in open areas.

When using this drag, keep in mind that pulley systems work best if the angle between the fixed and hauling lines is kept to a minimum. Ideally the firefighter(s) pulling the working end of the rope will be in line with the solid object securing the rope. In some cases the team may want to use their own search line instead of one specifically for hauling. Following the above instructions, the team turns the search rope back on itself after running the working end through the loop. They then pull together toward the exit, following the fixed end of their rope.

Teams likely will find this rescue technique highly effective if they use it correctly. Its primary drawback is that it is less effective going around corners, which are found throughout residential occupancies. (The friction of the rope against the corners adds resistance.) In large commercial areas, however, this drag is among the most proficient.[5]

Moving a Firefighter Up Stairs

Anyone who has attempted to move an unconscious firefighter up a flight of stairs can attest that it is not as easy as others might think. This is especially true on narrow residential stairs or with a large victim. Depending on the situation, the entry team may choose to have the second team deployed with lifting equipment or as a relief crew to take over dragging once the victim is at the top of the stairs. As with any structural fire-fighting operation involving stairs, rescuers should perform the following evolutions only after establishing the stairwell's structural integrity. Limit the load on wood stairwells as much as possible, and be sure to direct most of your body weight toward the stairwell's supporting members.[6]

Two-Firefighter Rescue

1. Using an appropriate drag, the rescuers move the victim to the base of the stairwell and place him in a sitting position with his back against the stairs (Figure 7.19).

2. One rescuer moves onto the stairs behind the victim and grasps the victim's SCBA straps. If the straps are difficult to locate or broken, webbing or a personal rope can be used instead. The rescuer then pulls the victim's body onto the stairs.

3. A second rescuer at the victim's feet now lifts the victim's legs over his (the rescuer's) shoulders, moves his head as close as possible to the victim's groin, and wraps his arms around the victim's legs to secure them (Figure 7.20). This position will provide excellent leverage and should be maintained throughout the stair rescue.

4. When both RIT members are in position, the firefighter at the head begins the operation by calling "lift." The rescuer at the feet pushes up with his legs to attain a standing position, while the rescuer at the head lifts the victim up to clear his SCBA from the steps. If the

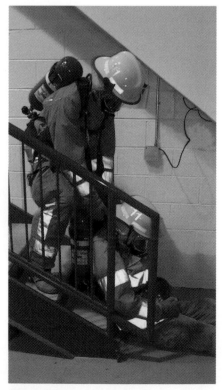

Figure 7.19 Position the victim with his back against the stairwell.

Figure 7.20 Positioning for the two-firefighter rescue up stairs.

victim is light, the rescuers will be able to remain standing and carry the victim right up the stairs. For heavier victims the team may need to repeat the lift sequence a number of times, moving one or two stairs with each repetition.

Two-Team Rope Rescue

The two-team rope rescue is similar to the two-firefighter method, but it is more useful for moving heavy victims or when interior conditions prohibit standing. This technique requires at least three or four rescuers.

1. The entry team positions the victim as in steps 1 and 2 of the two-firefighter method.

2. The second team is deployed to the top of the stairs with a rope bag. They retain one end of the rope and throw the bag (or one end of a rope with a pretied knot) down to the entry team.

3. The entry team secures the rope to the victim's SCBA harness.

4. As the second team pulls the rope from the top of the stairs, the entry team lifts the victim's head, shoulders, and SCBA cylinder clear of the stairs (Figure 7.21). When the victim reaches the top of the stairs the rope can be used to assist in pulling him from the building.

Rescuers can attach the rope to the victim's SCBA harness in several ways. One is to attach a carabiner to a figure-eight on a bight at the end of the rope. The carabiner will clip easily onto the harness or SCBA frame in the dark. Alternately, rescuers can feed the rope under the victim's SCBA shoulder straps and connect the rope back onto itself with a knot or, preferably, a carabiner on a figure-eight knot (Figure

Figure 7.22 An easy way to secure the hauling rope is through the victim's SCBA harness.

Figure 7.21 A second-team rescuer uses a rope to assist with hauling up stairs.

7.22). A third method is to pass the working end of the rope through the victim's SCBA straps and return it to the top of the stairs. This gives the rescuers at the top of the stairs two rope ends to pull and controls the unused bulk of rope, which remains in the bag.

Litter-Assisted Rescue

The previous two rescue techniques might not be effective for moving very heavy downed firefighters. The litter-assisted rescue uses a Stokes basket to slide the victim up the stairs. Setting it up takes slightly longer, but it ultimately requires far less RIT energy than the previous techniques. It is also the preferred method for removing firefighters with neck or spinal injuries.

1. As soon as possible upon determining the need to move a heavy or injured victim up stairs, the entry team should call the second team to bring a Stokes basket.

2. The second team, with one or two ropes and the Stokes basket, follows the search line to the top of the stairs.

3. While the entry team moves the victim to the bottom of the stairs, the second team attaches the rope(s) to the top of the basket.

4. The second team slides the basket down the stairs to the first team. The first team secures the victim in the Stokes basket and then lifts the top end of the basket onto the first couple of stairs. The second team pulls on the rope(s), sliding the victim up the stairs (Figure 7.23). The entry team assists by pushing the bottom of the basket.

If the entry team determines that the victim may have neck or spinal injuries, they can have the second team bring the basket directly to the victim. A backboard in the basket will provide immobilization and facilitate lifting. If a Stokes basket is not available, rescuers can use this same technique with only a backboard. The one inconvenience of using the board alone is that the victim must be strapped down (including straps under his feet) before removal, to prevent sliding off.

Placing a Firefighter in a Stokes Basket. Placing a downed firefighter into a Stokes basket under fire or smoke conditions can be difficult. If the victim is large, two members may not be able to lift him. Furthermore, most baskets are not large enough to safely carry the victim still wearing SCBA. While in some cases simply lifting the victim into the basket may be easier, teams should be fully trained in the following technique in case they encounter a difficult situation.

1. Remove the victim's SCBA harness. The easiest and safest way to do this is to release the SCBA waist belt and then cut one or both shoulder straps with paramedic shears. (Be careful not to cut the air supply line.) Alternately, slide the shoulder straps off the victim by turning his body. This may require more movement of injured areas and can take longer. In removing the harness assembly, be sure not to interrupt the victim's air supply.

2. If the victim's cylinder is low on air, provide a spare SCBA at this point. If the victim's air supply is adequate for removal, place his airpack into the basket and secure it on top of his body or between his legs in an inverted position. Keep the victim on his side after removing his air pack. Place the basket behind the victim and roll him onto his back and into the basket (Figure 7.24).

3. Fasten the basket's seatbelt-style straps to prevent the victim from sliding out. If the basket does not have straps of its own, connect clip straps or a rope before bringing the basket into the hazard area.

Using a Backboard with a Stokes Basket. To use a backboard with a Stokes basket, follow these steps:

1. Remove the victim's SCBA harness.

2. After removing the SCBA harness assembly, keep the victim on his side. Place the pack in front of his body and slide the long board in behind him.

Figure 7.23 The use of a rescue basket can greatly facilitate victim removal up stairs.

Figure 7.24 Placement of a firefighter into a rescue basket.

3. If possible, log roll the victim on to the board.

4. Each team member grasps one handle on the board. In a four-point lift, the rescuer's arms will keep the patient in place on the board. Tip the basket toward the victim and lower the board into it.

5. Secure the victim in the basket with straps.

6. Once the patient is secure, removal can be rapid, especially in larger commercial occupancies with wide hallways. Team members can easily push and pull the basket to the nearest exit, including those up or down stairs. As conditions improve, a four-point carry will further speed removal (Figure 7.25).

Obviously, immobilizing a downed firefighter generally will require more removal time than a simple drag or carry without apparatus, especially in residential occupancies. Drags and carries, however, are not the best ways to move a firefighter with a spinal injury. RITs must rapidly analyze the risks before moving any victim. Do the interior conditions indicate an immediate rescue, or is there time to take more care and call additional resources? Furthermore, if a number of obstacles such as staircases are to be negotiated, using a Stokes basket may ultimately decrease rescue time.

Moving a Firefighter Down Stairs

The easiest way to bring an unconscious firefighter down a narrow stairwell is a forward drag. The victim should be dragged to the top of the stairs, on his side, with his head at the edge of the top stair. The first rescuer grasps the top shoulder strap of the victim's SCBA and prepares to drag. The second rescuer moves down the stairs and stands below the first rescuer. This way, if the first rescuer loses his balance during the drag, the second

Figure 7.25 An injured firefighter can be efficiently removed using a three- or four-point backboard carry.

Figure 7.26 Firefighter removal down stairs.

rescuer will be there to steady him. When ready, the rescuers go down the stairs together, with the first rescuer pulling the victim by the SCBA strap. Under high heat conditions this drag can be performed in a crawling position (Figure 7.26).

For heavy victims on wider stairwells, the two rescuers can each grasp one SCBA shoulder strap and drag, facing either forward or backward. If they are using a Stokes basket or long board to move the victim, the RIT should pause at the top of narrow stairs. One rescuer should go ahead to check the stairs for stability and guide the basket. Using one or two ropes, the remaining team members slide the basket down the stairs. At the bottom, the RIT can either carry or push/pull the basket to the nearest exit or the next obstruction. If the stairwell is wide enough, the team may carry the basket or board, although sliding will expend less energy.

Firefighter Trapped Below the Area of Operation

Typically implemented only in desperate situations, "through-the-floor" evolutions are staged from a floor or roof above a firefighter who is trapped below the area of operation. The RIT team should attempt this dangerous technique only after assessing all other means of extrication, such as stairs, elevator shafts, and storm doors, and ensuring that the trapped firefighter is in immediate danger from fire, smoke, or other hazards. If no such danger is present, the team should use a rescue method that accommodates spinal injury.

If a through-the-floor rescue is necessary, the team must move quickly and efficiently. After sizing up the situation and ascertaining the need for immediate rescue, the RIT should perform one of the following two evolutions.

Figure 7.27 An RIT member uses a saw to widen a hole in the floor for firefighter rescue.

Figure 7.28 Accessing the victim via ladder.

Two-Rope Method. The team may use the two-rope method both for victims who are wearing an SCBA and for those who are not. Although three firefighters can perform this task, five are ideal.

1. The entry team calls for additional team members, two 50-foot rope bags, and a saw if necessary.

2. While waiting for backup, the first team assesses the surface area through which the trapped firefighter fell. The area around the hole must be sound enough to work from, and the hole itself should be large enough to remove the downed person.

3. Upon the second team's arrival, the crew enlarges the hole if necessary, being sure to cut no more than one structural beam (Figure 7.27). They may lay short ladders or interior doors across the remaining beams to provide a stable work platform. (Using a Rabbit Tool, one crew member can remove a door in as little as thirty seconds.)

4. One crew member either enters the hole by ladder (such as a closet ladder) or is lowered on one of the 50-foot rope lengths (Figure 7.28). Generally lowering is accomplished by tying a handcuff knot at the midpoint of the rope and placing it over the opening. The rescuer uses the handcuff knot as a foothold, while team members on either side lower him to the floor below. Once the rescuer reaches the victim, he places the handcuff knot around the victim's wrists or lower arms. The team members above then lower the standing part of the second rope, and the rescuer secures the victim's ankles with a handcuff knot at the rope's midpoint.

5. RIT members above now work in unison to raise the victim. As the rescuer below pushes up the victim's SCBA bottle and then feet, the two rope teams coordinate their efforts to ensure bringing the victim through the hole in the vertical position, wrists first. To do this, the team leader calls "lift" each time the team is ready to raise the victim further. After each lift the team should reassess their grip on the rope and their footing.

6. Once the downed firefighter reaches floor level, one RIT member grabs him by the gear or harness to facilitate lifting, and the team drags the victim forward out of the hole.[7]

Single-Rope Method. Although the two-rope method works well with a firefighter wearing an SCBA, the following method, created by the Fairfax County, Virginia, Fire Department, is arguably more efficient at raising victims who are wearing SCBA. This method uses just one rope and from three to five rescuers.

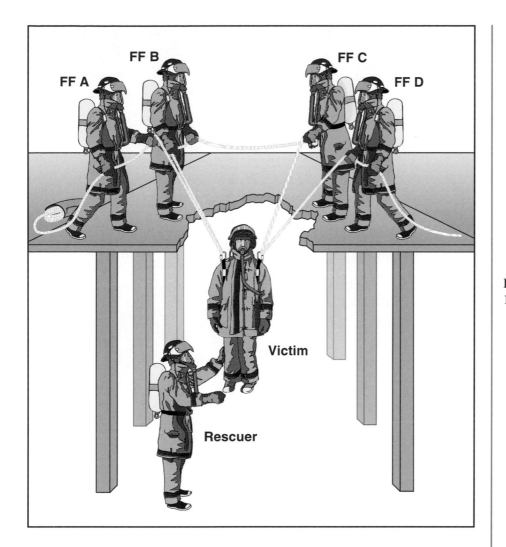

FF A FF B FF C FF D

Victim

Rescuer

Figure 7.29 RIT positioning for the 1-rope method.

The rope is prerigged with two figure-eight on a bight knots, one 10–20 feet in each direction from the rope's midpoint. (The ceiling heights in the region's occupancies should be considered when placing these knots.) A carabiner is added to each eight, and the rope is packed in its bag, ends first, until only the carabiners are left. Each carabiner is then clipped to one of the rope's handles for storage.

After the RIT has prepared the entry hole as in the two-rope method, one rescuer descends. Each of the four remaining rescuers stands at one of the four corners of the hole, one on either side of both prerigged knots. They lower the knots together, forming two equal loops in the rope. They remove only as much rope from the bag as needed to reach the victim. The rescuer in the hole then clips a carabiner to each of the victim's SCBA straps. Once the victim is secured, the team members on top pull him to the surface, while the rescuer below assists by pushing up on the victim's SCBA and/or feet (Figures 7.29, 7.30, 7.31). As in the two-rope method, the team should coordinate their efforts to maximize lifting efficiency. When the victim clears the edge of the hole, one rescuer can reach down and grasp the victim's SCBA straps and pull him to the side of the hole.[8]

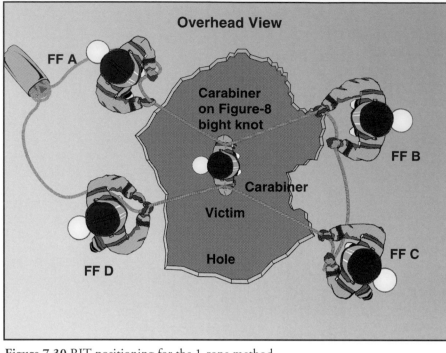

Figure 7.30 RIT positioning for the 1-rope method.

Figure 7.31 A rescuer assists with lifting from below.

Both of the "through-the-floor" evolutions should be modified according to the level of hazard present and the RIT manpower available. If only three people are on the team and the victim is light, two members on top may suffice. If the downed firefighter is wearing a personal harness, the RIT could raise him using a carabiner through the harness strap instead of the handcuff knot or SCBA straps. Determining company operating preferences regarding these techniques requires considerable practice.

Collapse Rescues

A structural collapse at a fire scene can be one of the most dramatic and stressful events a firefighter ever encounters. This is especially true if fellow firefighters become trapped in the debris. While even the most veteran RIT members may not have experienced a collapse with entrapment, the team must remain calm through the potential chaos.

RITs typically are not intended to fill the role of a collapse rescue team. If the team is trained and equipped for such incidents, all the better, but in many cases the RIT's primary responsibility will be ensuring that properly qualified help comes as soon as possible. Ideally, the RIT will recognize the signs of a potential collapse and call for backup before the collapse actually occurs (Figure 7.32). These warning signs include:

Figure 7.32 Heavy fire volume is a warning sign of potential collapse.

- Fire in or under truss construction components
- Cracks in walls, floors, ceilings, or chimneys
- Falling debris, loose bricks, stones, or blocks
- Cracks or deterioration in the building's mortar
- Walls that are leaning or twisting, or rumbling noises inside or outside the building
- Structural members pulling away from walls, floors, or ceilings
- Heavy machinery on upper fire floors
- Prolonged fire exposure in focused areas
- Reinforcement stars visible from the exterior
- Concrete breaking off or missing from rebar cages
- Column failure
- Floor displacement

If a collapse occurs, the RIT must determine if it can enter the hot zone. While this is a difficult decision, especially when a fellow firefighter is trapped, teams must consider their level of training before acting. Ideally RIT personnel will be qualified at least to the first level of collapse rescue readiness, *Basic Operations,* as established in the standards for collapse rescue training outlined in NFPA 1470. RIT members with this qualification have the requisite knowledge to recognize potential secondary collapse hazards and to rescue victims trapped in surface or minor debris.

Once a rapid intervention team decides to enter the hot zone, it must make certain preparations. First, the RIT must ensure that a collapse rescue team has been dispatched, even if its need is uncertain. This team can act as a second RIT if the original RIT team becomes trapped. Before the team touches any debris, the IC should conduct a personal accountability check to determine the number of victims. The RIT also should determine the specific locations of all victims if possible. Next, they must note any structural instability caused by the initial collapse and visually check for secondary collapse hazards that could follow the removal of debris. If such conditions exist the RIT should not enter. If the scene appears relatively uncomplicated, the team can continue its size-up by visually scanning for other hazards such as utility wires or pipes, slippery surfaces, and dangerous substances. Only after the team has addressed these concerns should they proceed to remove debris.

In many RIT situations the team can complete this size-up very quickly. For example, if the entry team locates a downed firefighter entrapped by a beam, a hands-on survey might find that the beam is no longer connected to the structure and can be moved safely. By ascertaining if the beam is connected to the structure, the team is in effect performing a secondary collapse survey. Hopefully, the team already will have considered certain other hazards such as uncontrolled utilities in its ongoing exterior size-ups. If, however, the RIT finds a complicated entrapment, the best course may be to wait for more specialized help before attempting a rescue (Figure 7.33). In the meantime, the RIT could place hand lines for victim protection, secure the downed firefighter's air supply, and establish scene lighting. Once the collapse rescue team arrives, the RIT should continue their support efforts under the direction of the more qualified team.

If the RIT encounters a particularly dramatic collapse, such as that of an exterior wall onto a group of firefighters, chances are that numerous firefighters may charge into the debris to save their comrades. At that point, most of the considerations and tactics described above become moot. The

Figure 7.33 Structural collapse rescues may require specialized RIT resources.

RIT must decide how it can best protect as many lives as possible, including their own. Among a few options the team has are:

- Standing outside of the collapse zone and calling additional resources

- Blocking entry to areas of greatest hazard to prevent further injuries

- Performing a size-up from a distance

- Attempting to block the entire area and limit access to those with proper training

There are no easy explanations for how RIT members should act in a massive collapse involving firefighters or other emergency personnel. Obviously, the best way to prepare for such situations is to train all department members in structural collapse rescue awareness *before* disaster strikes. If firefighters understand the hazards and potential counterproductivity of charging into a collapse zone, they will be less likely to do so.

Conclusion

As with any rescue, the easiest and safest method of extrication should be used to remove a downed firefighter from a building. RIT members must try to consider all their options before moving a victim. They should base their decision on a number of factors, including interior conditions, the victim's medical situation, and the availability of resources. Team members must always perform an ongoing "cost-benefit" analysis to determine their best rescue option. For example, although dragging a victim with a spinal injury generally is not advisable, it may be a "cost" of gaining the greater "benefit" of removing the firefighter from a burning room. Generally speaking, costs should never exceed benefits.

While the techniques in this chapter do not address every complex rescue situation that could be encountered on the fire ground, they can equip RIT members with a basic mental toolbox of interior evolutions. Individual departments and companies should further develop these techniques and adapt them to specific local scenarios. Frequent company drills, training courses, and discussion will ensure that department members continue to add to and upgrade their rescue skills throughout their careers.

Notes

1. In addition to material specifically developed for this book, this chapter and the next include evolutions from fire departments and training facilities across North America. All published techniques are referenced to their earliest publication, according to a bibliographical search performed at the National Fire Academy.

2. This technique appears in John Norman and John A. Jonas, "Unconscious Firefighter Removal," FDNY training bulletin, undated.

3. See Kieran J. Ordway, "Fallen Firefighter Drag Rescue," *Fire Engineering*, August 1999, pp. 54–56, and Art Donahue, "RIT Rope Drag," *Fire Engineering*, February 1999, p. 14.

4. This drag was taught during the "Saving Our Own" session at FDIC's 1997 "HOT" program. It also appears in Norman and Jonas, "Unconscious Firefighter Removal."

5. The 2:1 pulley drag is explained in Norman and Jonas, "Unconscious Firefighter Removal."

6. Variations of the stair techniques from Rick Lasky's "Saving Our Own" program developed at the Illinois Fire Service Institute were taught at FDIC 1997 and were later published by Rick Lasky and Tom Shervino as "Saving Our Own: Moving the Downed Firefighter up a Stairwell," *Fire Engineering*, December 1997, pp. 14–18.

7. This method is adapted from Rick Lasky and Ray Hoff, "Saving Our Own: The Firefighter Who Has Fallen through the Floor," *Fire Engineering*, March 1998, pp. 12–18; and Norman and Jonas, "Unconscious Firefighter Removal."

8. See Robert Dubé, "Rescue Rope for Rapid Intervention Teams," *Fire Engineering*, January 2000, p. 14.

Step 1: After forming a bight in the working end of the rope; pass the bight over the standing part of the rope to create a loop.

Step 2: Pass the bight under the standing part and through the loop.

Step 3: Dress the knot by pulling the bight and standing end away from each other.

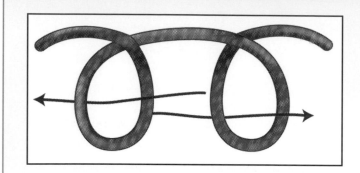

Step 1: Form two loops anywhere in a length of rope, as shown.

Step 2: Simultaneously pull the inner side of each loop through the opposite loop to form two new loops.

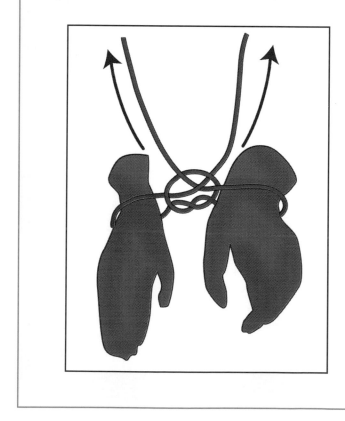

Step 3: Place the knot over the victims hands or feet and then tighten by pulling the two rope ends away from one another.

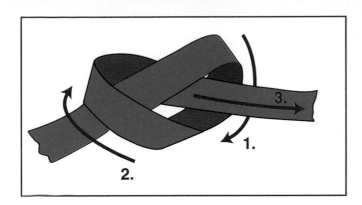

Step 1: Tie a simple overhand knot loosely at one end of a sling or personal webbing.

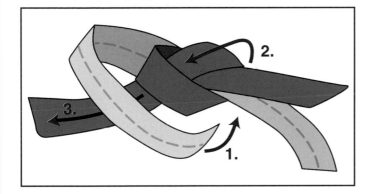

Step 2: Trace the other end of the sling/ webbing (shown in yellow) through the overhand knot, moving from the working end to the standing end.

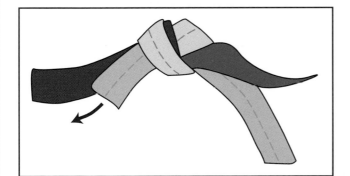

Step 3: Pull on each end of the webbing to tighten the knot and then tie a simple overhand knot at each end for safety.

Chapter 8
Exterior Rescue Evolutions

Exterior Rescue Evolutions

Although the RIT can accomplish many firefighter rescues by dragging or carrying the victim through the fire building, this may not always be possible. Consider a scenario in which one or more heavy victims are located in a third-floor rear bedroom. Heavy smoke and fire conditions exist on the first and second floors, and the downed firefighters have limited air. The safest and most efficient means of rescue is not through the house but out the bedroom window. Instead of calling the second team into the building, the entry team has them prepare for removal from the outside via ladders or ropes (Figures 8.1, 8.2).

The evolutions in this chapter involve deploying the second team from the outside. While the exterior team sets up, the first team prepares the victims and secures the room they are in by closing doors and performing ventilation. For training purposes, a designated safety officer should oversee these evolutions. Due to the hazard of falling, redundant safety ropes should be in place in addition to those used for the firefighter rescue.

Moving a Firefighter into a Window

As with many other RIT techniques, the effort required to move a downed firefighter into a window is often underestimated. The main methods for placing a victim in a window are face-up, face-down, and flip-over. They are prerequisite to most of the evolutions in this chapter.

Face-Up Method

The face-up method is very useful for moving a firefighter face-up onto a ladder or for removal via a simple rope rescue basket. First, drag the victim to the window area using a method suitable to the interior conditions and the firefighter's size. Next, place the victim directly under the windowsill with his legs pointing toward the wall. Lift his legs onto the windowsill and

Figure 8.1 Heavy fire on lower floors may require RITs to access victims via windows.

Figure 8.2 Debris may require RITs to access victims via windows.

maneuver his body toward the window until his buttocks are touching the wall or ledge (Figure 8.3). One rescuer kneels on either side of the victim and cradles him under the buttocks and arms. If the victim is still wearing an SCBA, the straps can assist with lifting. Using the windowsill as a fulcrum, pivot the victim up and forward to a sitting position in the window.

Note: Due to the obvious falling hazard, make certain the receiving ladder or rope system is in place before lifting the victim into the window.

Face-Down Method

The face-down method, which slides the victim onto a ladder feet first, is useful for heavy victims. Two RIT members can perform it, but three work better. To begin, lay the victim on his back *at one side of the window* with his feet against the wall. One rescuer lies on his or her stomach *in front* of the window with his feet also against the wall. This rescuer will have to remove his SCBA harness and place it beside him, on the side away from the victim. The remaining RIT personnel roll the victim onto the rescuer's back. When the victim is in place, the first rescuer pushes his body up into a crawling position. The remaining team member(s) guides the victim's legs into the window and to an exterior team member on the ladder.[1]

Alternatively, rescuers can follow the steps for the face-forward method, but turn the firefighter in the window. This can be done fairly easily if a second team member on a ladder outside the window assists the entry team. The rescuer on the ladder rolls the victim's legs around while the interior team rolls his body on the sill, using the SCBA harness as handles (Figure 8.4). This method works well with most windows except those that are very small.

Figure 8.3 Preparing for the face-up method.

Figure 8.4 Turning a victim face-down in the window.

Flip-Over Method

The flip-over method of placing a firefighter in a window may appear unorthodox, but it can be highly effective. Begin by dragging the victim headfirst to the windowsill. With his head near the wall, grasp the firefighter's legs, pull them together over his head, and then maneuver them toward the windowsill (Figure 8.5). An RIT member outside the window can grasp the feet as the rescuers inside roll around the rest of

Figure 8.5 Positioning for the flip-over method.

Figure 8.6 Flipping the victim into the window.

the victim's body. The victim's body will somersault into a face-down position in the window (Figure 8.6). This method's advantages include its speed and the fact that the victim's SCBA can remain in place. Its disadvantages are that the team may break the victim's SCBA mask seal as they roll him or place harmful pressure on his neck if they use an incorrect technique.

Moving an Unconscious Firefighter Down a Ladder

Perhaps the safest method of rescuing a firefighter from an upper window under hazardous conditions is via ground ladder with assistance from outside personnel. Victims can be brought out face-up or face-down, depending on team preference and interior conditions.

Forward-Facing Method

The forward-facing method calls for the second team to assist with the rescue from outside.

1. If possible, move the downed firefighter to a safe room. Drag him to a suitable window or other egress point, and place his feet against

Figure 8.7 Placing a firefighter in the forward position on a ladder.

Figure 8.8 Bringing a victim down a ladder in the forward position.

the wall directly below the opening. The victim should be either on his back or side, facing upward.

2. While the entry team removes the downed firefighter's SCBA harness and places it beside him, the second team places a portable ground ladder, and one team member climbs to the window.

3. The entry team places the victim onto the sill using the face-up method, and the rescuer on the ladder receives him (Figure 8.7). The victim's SCBA regulator should be left plugged in until the last possible moment. Depending on his size and the length of his SCBA low-pressure hose, the interior team may need to do this before lifting.

4. The entry team places the victim's legs over the ladder rescuer's shoulders. The rescuer then steps down one or two rungs and grasps the rungs under the victim's armpits while the entry team holds the victim in place. When ready, the exterior rescuer climbs down the ladder, placing his hands firmly on alternating rungs to support the victim (Figure 8.8).

Inward-Facing Method

The inward-facing method is very similar to the forward-facing method, but the victim faces down throughout the rescue. This allows his SCBA to remain on the entire time, but it makes the climb down the ladder slightly more difficult for the exterior rescuer.

1. If possible, move the downed firefighter to a safe room. While the exterior team places a ground ladder in the window, the interior team uses either the face-down or flip-over method to place the victim in the sill (Figure 8.9).

2. Once an exterior team member has climbed to the window, the interior team guides the victim's legs out to the exterior rescuer.

3. As the interior team moves the victim out, the exterior rescuer guides the victim's legs along the beams until the victim's body rests on the rescuer's knee. The victim's armpits should rest on the rescuer's forearms, with the rescuer grasping the rungs underneath (Figure 8.10).

Figure 8.9 Placing a victim for an inward ladder removal.

Figure 8.10 Lowering a victim using the inward ladder method.

4. As the rescuer descends the ladder, he shifts the victim's weight from one leg to the other, just as he would typically do a civilian rescue. The placement of the rescuer's arms helps to ensure that the victim's SCBA regulator stays in place.

When using either of these ladder rescues, the RIT can exit safely via the ladder after the victim's removal.

Ladder Rope Rescues

Ladder rope rescues, which use the second team for exterior support, are effective for removing heavy victims from floors above grade or from a small room with a window.[2] Rescuers should use these rope methods if they cannot safely and efficiently remove a victim through the fire building. Rescuers must remember that single rope rescues under fire conditions can be hazardous.

Basic One-Rope Method

1. The entry team assesses its location and orders the exterior team to set up outside the closest window. If possible, the entry team moves the victim to a safe room, as necessary.

2. While the exterior team positions a ground ladder with its tip at least two rungs above the *top* of the window, the interior team prepares the victim for removal by converting his SCBA into a harness and then sitting him with his back against the windowsill. One way of doing this is to drag the victim headfirst to the wall directly under the window. One rescuer moves to the victim's feet and pulls him by his arms to a sitting position. Together the team slides the victim back against the wall under the window in a sitting position (Figure 8.11).

3. Once the ladder is in place, one member of the exterior team carries a rope bag to the top while another exterior member (such as the RIT driver) foots the ladder for the remainder of the evolution. As with all nonspecialty (not prerigged) RIT ropes, this rescue line should have a loop and carabiner at each end. The ladder rescuer passes its working end into the interior team members, who grasp it tightly. The ladder rescuer then feeds the bag over a top rung and drops it to the ground. The ladder rescuer then descends as the team member on the ground flakes out the rope.

4. The interior team slips the working end of the rope through the victim's SCBA straps and secures it with the carabiner (Figure 8.12).

5. While the exterior team takes up the slack on the rope and pulls to raise the victim out of the room, the interior rescuers assist with lifting the victim clear of the window. Once the victim is

out, the exterior team lowers him to themselves on the ground (Figure 8.13).

This rescue can be modified in a number of ways, depending on local preference. First, the team may choose to lower the victim using any number of rescue harnesses, either manufactured or tied from rope or webbing (Figures 8.14, 8.15). Rescuers considering the lowering method must weigh the hazards of the scene against the time required to attach a

Figure 8.11 Victim placement for the basic one-rope method.

Figure 8.12 A rope with carabiner through the SCBA straps for lowering.

Figure 8.13 Basic one rope method seen from the exterior.

Figure 8.14 and 8-15 Any number of rescue harness variations can be improvised on the fireground.

harness other than the SCBA. A variation of this rescue is to feed the rope under the bottom rung of the ladder. Although this placement's added friction makes *lifting* the downed firefighter more difficult, it also makes *lowering* him easier and safer.

2:1 Pulley Rescue

The 2:1 pulley rescue technique is very similar to the basic one-rope method, except that it incorporates a prerigged 2:1 pulley system to overcome lifting friction caused by the ladder rungs.

To prepare the pulley system, follow these steps:

1. Pack 100 feet of life-safety rescue rope in a suitable rope bag.

2. Connect a hook through the top of a single pulley to secure it in a closed position.

3. Run the working end of the rope through the pulley and tie a carabiner/hook to the end of the rope. (This prevents the rope from slipping back through the pulley.)

4. Clip the pulley's hook onto the outside of the bag for easy access and slip the working end of the rope into the bag.

5. Tighten the bag for storage on the RIT vehicle.

To begin the 2:1 pulley rescue, the RIT should complete steps 1 and 2 of the basic one-rope method. A member of the exterior team then carries the prerigged pulley system up the ladder (Figure 8.16). Next, the ladder rescuer clips the pulley's carabiner onto the ladder's first or second rung. (To prevent the pulley from sliding across the rung, rescuers can use webbing tied in a loop with a water knot as an anchor point for the pulley carabiner (Figure 8.17).) The ladder rescuer hands the working end of the rope in to the entry team, as in the basic one-rope method, and then drops the rope bag *directly* to the ground (Figure 8.18). As soon as the entry team has secured the victim, members of the exterior team pull down on the rope to raise the downed firefighter out of the room and then lower him to the ground.

Alternatively, Mark McLees of the Syracuse, New York, Fire Department suggests lowering the pulley itself to the interior rescuers. To do this, the ladder rescuer attaches the carabiner at the working end of the rope to the top rung and hands the pulley through the second rung to the interior team. He then drops the rope bag between the second and third rungs to the ground (Figure 8.19).[3]

Figure 8.16 The use of a prerigged pulley.

Figure 8.17 The carabiner can be attached directly to a ladder rung or to a sling attached to the ladder.

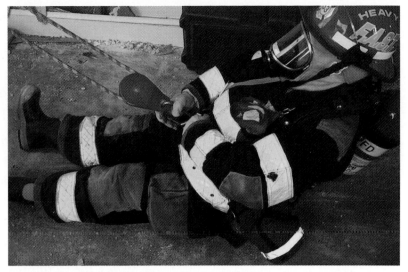

Figure 8.19 An alternate rigging for the pulley rescue.

Figure 8.18 The remainder of the rope is dropped to the ground.

Using Block and Tackle

If the RIT must remove an extremely heavy victim through a window, they may consider a double block and tackle. Greatly reducing the downed firefighter's weight but highly prone to tangling, such systems should be kept prerigged on the truck and used only if necessary. Before deploying the block and tackle, team members must assure that the rope used to rig the system will be long enough for the rescue once the blocks are separated.

Rescues using a block and tackle follow the steps for the 2:1 pulley rescue, with one main difference. Once the ladder rescuer secures the standing block to the ladder, he must lower the running block into the room with the victim. The interior team can attach the large hook on the bottom of the running block to the victim. They then should carefully guide the firefighter out the window to avoid twisting the rescue rope.

Simple Rope Rescue out a Window

Simple rope rescue out a window is an easy way to lower a firefighter from an upper floor to the ground when ladders are not available or feasible. For example, the window may be too high up or the ground around the building may be too uneven. It can be utilized with a small to medium-sized victim at commercial, residential, and even high-rise occupancies.

1. If possible, the entry team moves the downed firefighter to a safe room. They drag him to a suitable window or other egress point and place his feet against the wall directly below the opening, on either his back or his side, facing upward.

2. The entry team calls the second team for a life-safety-rated rope and a tag line, both long enough to safely reach the ground. (Allow at least 12 feet per floor and 50 feet excess for hauling.) While waiting for the second team, one member of the entry team completely breaks out the window. The RIT member remaining outside the building moves to a position where he can see the window.

3. When the second team arrives at the victim, they secure the rope to him. Some crews may simply slide the rope under the victim's SCBA straps after converting the pack to a harness; however, a pretied rescue harness is much safer. As always, the entry team must conduct an ongoing risk analysis and ask themselves if this is the best method of removal, given the hazards present.

4. Once the victim is secured to the rescue rope, the rescuers should clip the utility line to the front of his body and throw it to the firefighter on the ground as a tag line. The entry team members flake out the rescue rope toward the rear of the room and grasp it with their gloves while the second team positions the victim in the windowsill using the face-up method.

5. When ready, the entry team takes up the slack from the rope, and the second team members each grasp one side of the victim's SCBA or rescue harness. The second team lowers the victim out the window as far as possible by allowing his buttocks to slide off the windowsill; they continue to support some of the victim's weight by maintaining their hold on the shoulder straps.

6. Once the entry team ensures they have a firm grip on the rope, the second team lets go of the victim. The firefighter on the ground applies light pressure to the tag line to keep the victim off the building's wall. One second-team member grasps the rope in front of the entry team; the other stays at the sill to brake the rescue rope by pressing down on it just behind the sill. This team member also keeps visual contact with the rescuer on the ground. To lower the victim, the firefighters on the rope simply allow the line to slide through their hands. To slow the rope they tighten their grip.

7. When the victim reaches the ground, the firefighter there disconnects him from the rope and helps transfer him to EMS (Figure 8.20). If the rescuers need to lower a second victim, they can easily and quickly pull the rope back to the room and reuse it.

This evolution's greatest asset is its ease. Although explaining it takes a fair amount of time, in reality it can be done very quickly and uses just two ropes. As noted, it is less useful for especially large victims. If the downed firefighter is too heavy for the rescuers to lift into the window easily, additional personnel or a different rescue technique should be considered.

Figure 8.20 Rescuers on the ground must move to quickly release the victim from the rope harness and turn him over to EMS.

Ladder Slide

The ladder slide technique is designed to remove a firefighter with a back or neck injury from a floor above grade. It requires the victim to be in or near a room with a window.

1. Having determined that the victim may have a back injury, the entry team should move him only as much as indicated by the hazards present. If the victim is in a room that the entry team can secure by closing doors or venting windows, they should do so.

2. Once the victim is in an area free of immediate hazards, the second team should be deployed. The easiest way for them to access the victim is via ground ladder through the window to be used for the rescue. They should bring a long board and prepacked strap/collar bag, as well as a life-safety rope.

3. While the second team makes its initial entry, the team member remaining on the ground should prerig a Stokes basket with two tag lines, one on each of its bottom corners.

4. If time and conditions permit, when the second entry team reaches the victim, three of the rescuers can remove his SCBA and immobilize him. The remaining team member inside tosses the rope down to the team member on the ground.

5. The RIT member on the ground attaches the rope to the top of the prerigged Stokes basket, and the second team hauls the Stokes up the ladder. The fifth team member and an additional helper guide

the tag lines. If no additional helper is available, the firefighter at the top of the ladder can climb down, leaving the hauling line at the top. Another team member will pull in the basket.

6. Once the basket is in the room and the victim is immobilized, the RIT lifts the victim into the Stokes and secures him, placing his SCBA on top of his body. The rescuers then lift the Stokes into the window. As soon as the basket is on the sill, the rescuers who lifted the basket's foot into the window move to the rear and grasp the hauling line. Together with the rescuers lifting the basket's head, they pivot the Stokes onto the ladder. The rescuers at the head move to the rope, and the whole team controls the rate of descent. Even with a heavy firefighter, the friction of the rope passing through the rescuers' gloves and over the windowsill should be sufficient to control the basket's speed.

This is a highly effective rescue to remove any size victim with a back injury from a floor above grade. Throughout the process of securing the rescue basket, the RIT should monitor fire conditions carefully. If the fire goes out and air quality improves, simply carrying the victim out may be easier than using a ladder. Due to the large amount of equipment and the teamwork involved, RITs should practice this evolution regularly.

Aerial-Assisted Rescues

RITs should not discount the value of an aerial ladder to perform firefighter rescues. Because these units are typically placed in front of the fire building and reach far higher than ground ladders, they provide the team with an important tool. Upon arrival at the scene, the team should determine the locations of all aerials and determine whether an operator is with each vehicle. Ideally, the RIT driver should be trained to operate such equipment in the event that the units have no operators.

If a scenario arises where an aerial ladder will be utilized, the RIT officer will need to decide whether to send the entry team in through the building or via the ladder, whichever is most direct. Once they reach the victim, the entry team's options involving the aerial piece are limited only by their imagination and the need for safety. Certainly, the RIT can perform a number of rope based rescues using the aerial ladder as a base. They can also remove downed firefighters directly to the aerial ladder and down to the ground. Additionally, ladder slide operations can be greatly facilitated by a ladder truck's main (Figure 8.21).

Breaching Walls

Wall breaching is simply the creation or enlargement of holes in walls for the purpose of firefighter entry or egress. Such openings are often easy to make and give the RIT a rescue means that may be simpler than bringing the victim through heavy fire conditions or a complicated floor plan littered with debris. This is especially true if the rescuers are moving the victim by long board or Stokes basket or if the fire building has few doors (Figure 8.22). If the entry team chooses to use a wall breach, they must communicate their decision as soon as possible. Once the entry team locates the victim, outside and inside teams can work together to determine the best breach option. As the entry team prepares the victim, the exterior team makes the opening.

RITs generally must breach one of four types of wall construction: wood, masonry, metal, or gypsum board. Efficiently breaching each of these types of walls requires specific tools and techniques. Before beginning any breach, the team should first determine whether it will cause structural instability. Holes in load-bearing walls must be kept small and should avoid damaging framing studs if possible.

WARNING

Always wear full personal protective equipment and eye protection when breaching walls. Before cutting, ensure that the power to the building has been turned off.

Figure 8.21 Aerial ladders provide excellent access to upper floors.

Figure 8.22 The removal of a wall can facilitate rescue of a large or injured firefighter from a cramped space.

Wooden Walls

Wood is one of the most common types of building materials on the fire ground. It is also one of the easiest to breach (Figure 8.23). One of the most efficient ways to breach any type of construction is to start at an existing opening. For example, rescuers might find a window near the victim and cut the sill down to floor level. This will allow the easy passage of a boarded victim to the exterior. This evolution can also be performed to make the ladder slide rescue easier. If no windows or other openings are available, the outside team should first make a probe hole in the wall. After they determine support beam locations, they can make an appropriate opening.

The best tool for breaching a wooden wall is a power saw. After determining the size and location of the hole, the outside team should start the saw and make sure the chain or blade is turning at its maximum RPM before beginning the cut. As with a ventilation opening, they should not force the blade through the wood but allow the saw to "do the cutting" (Figure 8.24). Once the outline of the hole is made with the saw, a second firefighter should use a hook to pull away the materials and clear debris (Figure 8.25). If the saw has not penetrated all building materials, a second cut may be necessary.

Masonry Walls

Masonry walls include those made of brick, concrete, or stone. They are found mainly on the exterior of many homes and commercial buildings. Although not as easy to breach as wood walls, masonry walls can be a viable RIT egress point if the correct tools are available and if structural integrity is sound (Figure 8.26).

Figure 8.23 Sheathing is installed on the outside of exterior wood-frame walls to provide structural stability, weatherproofing, and to serve as an underlayer for the exterior finish material.

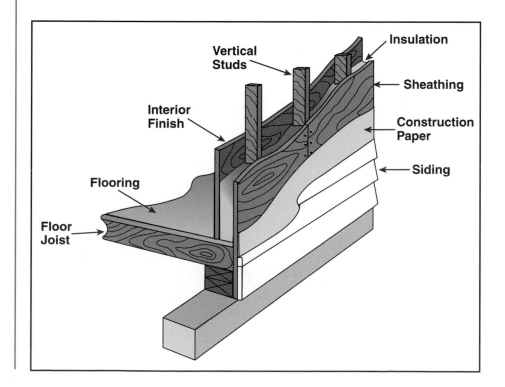

Figure 8.24 Enlarging a wall opening with saws.

Figure 8.25 Clearing debris from the breach site.

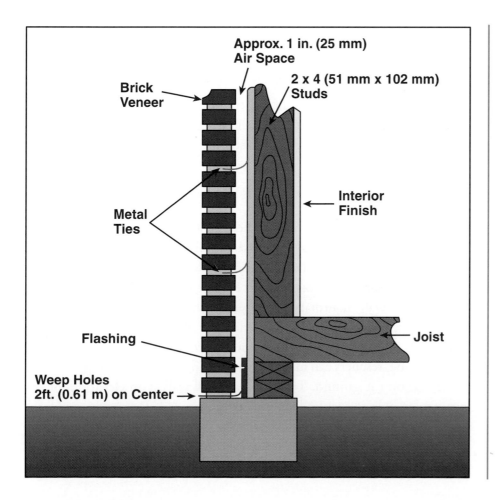

Figure 8.26 A wood-frame building may be faced with a brick exterior called a "brick veneer."

Figure 8.27 Sledgehammers and battering rams are highly useful for breaching masonry walls.

As with wood walls, the easiest way to perform a masonry breach is by enlarging an already existent opening. Sledge hammers are especially useful for this (Figure 8.27). If no opening already exists, the team will need to begin by creating an initial breach. They can use a large battering ram with a pointed tip to do this effectively. Once the team makes the initial breach, they can use sledge hammers, air chisels, or other power tools to widen it. When working with masonry, RITs should keep in mind that a diamond-shaped hole provides more structural stability than a square or rectangular opening.

Metal Walls

Metal walls are often found in commercial structures and mobile homes. Residential and commercial garage doors also fall under this category. Before breaching metal walls, rescuers should ensure structural stability. One or two probe holes may be necessary. The recommended tool for cutting metal walls is a rotary power saw with a metal-cutting blade. The method for smaller holes and garage doors at ground level is to make two long diagonal cuts from a common point down to the ground. Once these cuts are made, rescuers can make a triangular opening by folding the metal down, flat on the ground. To breach a metal wall at an elevation, or if a rectangular hole is required, four cuts are needed. Rescuers should cut support beams and studs only if necessary. After they make the cuts, they should carefully remove the metal plate and surrounding insulation and place it safely aside before attempting rescue.

Plasterboard Walls

Plasterboard walls (also called gypsum or drywall) are generally used as interior partitions and are the easiest walls to breach. From an RIT perspective, breaching these walls is most useful when a firefighter is trapped in a small room or closet. Perhaps the room's door is blocked by fire or a collapse has occurred. Breaching an interior wall typically takes less than a minute, generally does not hurt structural integrity, and requires only hand tools.

To breach a plasterboard wall, rescuers begin by determining the opening's location. The most difficult aspect of this choice is that available imaging technology cannot yet indicate if hazards or obstacles exist on the other side of a wall. (Thermal imagers can, however, indicate if there is fire *in* the wall.) The rescuers must make a probe hole to ensure that they do not make the breach in front of a piano or other heavy object. To do this, they should ram the pry-end of a Halligan bar through the plasterboard. If it meets no obstructions they can then turn the bar around and ram the adz into the probe hole, through the first layer of plasterboard. They then turn the adz inside the wall and pull out and down, removing large chunks of plasterboard (Figure 8.28).

Figure 8.28 Plasterboard is best breached using the adz end of a Halligan bar.

To enter a room through an interior wall breach without cutting the studs, the rescuer must turn his body at key times as he crawls between the studs. He should begin on his side and place both arms through the opening. Once his shoulders are in the room, he turns to a prone position to allow his SCBA to clear the studs and then pulls the rest of his torso through. If necessary, he turns back on his side to clear his pelvis once his SCBA is through the opening. This method is quicker and safer than removing his SCBA.

If the rescuers must move the victim through the hole, they likely will need to cut one of the wall studs. They can do this easily with a saw, which the second team brings in.

Breaching should be considered as one more rapid intervention tool. It can make certain difficult rescues far simpler by removing the need to pass through interior obstacles or hazards. It can be used for floors, stairwells, and roofs, as well as walls. Rescuers must, however, approach it with the correct tools and must always consider structural integrity before making entry.[4]

Other Exterior Team Support Roles

The second team's role is not limited to direct rescue functions in support of the entry team. They may have to fill engine or truck company roles in fire suppression. Such assistance can take many forms, ranging from hose line advancement to roof ventilation.

In the case of heavy fire conditions, entry to the victim may not be immediately possible. The second team can stretch hose lines to give the entry team an opening to access and rescue the downed firefighter(s). To be ready for such an operation, the RIT should ensure the availability of a viable water supply and reserve hose lines during the initial staging of equipment and throughout the incident. Readiness may be increased by pulling one or two spare hand lines from nearby engines. Andrew Fredericks of the New York Fire Department argues convincingly in favor of having a rapid intervention company engine (RICE) at each working fire, in addition to the regular RIT (Figure 8.29).[5]

Figure 8.29

Conclusion

This and the previous chapter provide numerous evolutions for the firefighter's "rescue toolbox." These techniques cover a wide variety of both interior and exterior assisted extrication scenarios that rescuers may encounter during working fires or even in training activities. These chapters do not, however, come close to covering every entrapment situation that could arise on the fire ground. Continuing drill and experimentation by RITs is vital. In many cases teams may combine several techniques to achieve a unique goal.

Regardless of how they achieve the rescue, team members must perform an ongoing risk analysis of their actions, continually asking themselves, "Is this technique the safest and easiest way to get this person out?" Although RITs should always attempt to act in the safest manner possible, certain fire situations may necessitate some risks when a fellow firefighter might die if not quickly extricated. Teams must always be prepared to scale down rescue techniques if fire or hazard conditions lessen. When firefighters' lives are on the line, RITs cannot afford to waste time.

Notes

1. This technique appears in John Norman and John A. Jonas, "Unconscious Firefighter Removal," FDNY training bulletin, undated.

2. For further discussion of such events and possible solutions, see Rick Laskey and Sal Marchese, "Saving Our Own: Removing a Downed Firefighter from a Tight Room and Window," *Fire Engineering*, April 1998, pp. 12–20.

3. See Mark McLees, "The Rapid Intervention Rope Bag," *Firehouse*, April 2000, pp. 40–42.

4. For further information on wall breaching see Gene Carlson, ed., *Forcible Entry*, 7th ed., Stillwater, Oklahoma: Fire Protection Publications, 1987.

5. For a more detailed discussion of engine company RITs, see Andrew Frederick, "Engine Company Support of RIT/FAST Operations," *Fire Engineering*, April 1999, pp. 79–96.

Chapter 9
Maximizing Use of RIT

Maximizing Use of RIT

Many small to medium-sized departments may find it difficult to establish an RIT from within their own ranks. It is hard to take good firefighters and make them stand by at working incidents, watching the situation unfold. Firefighters must look at RIT duty from a different perspective. The unit may be able to maximize its presence on the emergency scene by using its equipment to perform support duties that may be neglected due to other priorities and personnel shortages.

Rit as a Safety-Officer Function

As previously discussed, the RIT may best fit within the incident management system as a function of the safety officer. In this role, the RIT may be able to perform a number of additional duties.

Accountability. The RIT can help keep track of who is doing what on the fire ground, especially which units and people are operating inside the hazard area. Not only does this better inform the RIT of the whereabouts of personnel for whom they might have to search if something goes wrong, but it also fills NFPA fire ground requirements. This job may become quite taxing at larger incidents with numerous companies operating.

Access Control to the Hazard Area. Controlling access to the hazard area fits closely with accountability. Limiting the responders in the hazard zone to those who actually need to be there will provide for better accountability on the scene. This job may be difficult for an RIT in large buildings with multiple ingress points, but this is an area where the American fire service could make significant improvement.

Establishing Secondary Means of Egress. Methods of establishing secondary means of egress vary widely among different fire departments. Some departments are quite good at it, while others operate with such small crews that they can never even consider it. At a working incident, where construction, collapse, or fire spread may limit means of egress, secondary exits must be available. The RIT team can provide them by throwing ground ladders to the roof or upper floor windows (Figures 9.1, 9.2). This may provide a firefighter in distress an easy egress point.

The rapid intervention team can also provide secondary egress points simply by opening doors and windows that are protected with various security measures. They can swiftly and effectively accomplish this with power saws and small hand-operated hydraulic tools. The openings do not need to be pretty, but they do need to allow firefighters to exit the hazard area rapidly or to be rescued from the building should a sudden, unexpected event occur. A lock forced or a bar cut from a window can allow a disoriented firefighter to exit the building before his air supply is depleted (Figures 9.3, 9.4).

Lighting. Firefighters usually forget about lighting until the bulk of the fire is knocked down and everyone suddenly realizes that they are working in the dark. Then, they put helmet-mounted or handheld

Figures 9.1 and 9.2 RIT Placement of ground ladders provides additional egress points for upper floors.

Figures 9.3 and 9.4 RITs need to take action to ensure windows and doors are not barred preventing egress.

Maximizing Use of RIT

flashlights to work, but these are usually insufficient to clearly reveal the overall hazards on the fire ground. In these conditions, having a vehicle with power generation and lighting capabilities can be a strong advantage to a RIT. Being able to place a light tower in service or string tripod lights to the sides or rear of a building can significantly enhance safety on the fire ground (Figure 9.5). Even running a power junction box to a doorway and placing quartz lights at the egress point can enhance a crew's ability to see what they are doing and to leave the building quickly if need be. Strobe lighting or other types of beacons can be placed at building egress points to accomplish a similar objective. Departments should experiment with these techniques, as powerful strobes may tend to reflect off of heavy smoke, actually adding to the disorientation of personnel operating inside. All mutual aid units must use standard equipment and lighting signals so that everyone on the fire ground knows the meaning of particular lighting signals.

Figure 9-5 Scene lighting adds greatly to fireground safety.

Monitoring Changing Fire Conditions. The incident commander or various sector officers and the safety officer are responsible for monitoring changing fire conditions. They may or may not do this effectively on every fire ground. Another set or two of trained eyes monitoring the situation could save a life.

Rope Off/Demarcate Hazard Areas. The best way to handle a firefighter injury is to prevent it from ever happening. Any number of hazards can present themselves at a working fire, including live electrical wires, swimming pools, excavations, collapse zones, and others. With relatively minimal effort, the RIT can rope or tape off these hazard zones. This could prevent a fatigued firefighter from exiting the building with a fogged SCBA mask and walking into the hazard.

Controlling Utilities. While controlling utilities is traditionally a function of the truck company, this job may be forgotten early in the incident when many other tasks must be handled. Controlling electric or gas service to a building may prevent injury to those operating inside. Controlling water service may minimize collateral damage. At the least, the RIT officer can tactfully remind the IC that the utility companies should respond to the scene if they have not already done so.

Setting Up/Performing Rehab. Rehab is a topic unto itself. If properly performed, it can maximize the performance of the personnel on the scene while minimizing their health risks, particularly cardiac or stroke-related. These health risks are a real, significant cause of firefighter casualties. This task will most likely fall to the safety or EMS sector; however, carrying water, towels, and other basic supplies necessary to establish the rehab area can be an excellent, low-impact job for the RIT. Rehab is usually located near the RIT staging area anyway, and RIT members' medical training is likely to benefit this fire-ground function.

RIT members need to be careful to not get too involved in medical monitoring, but if they are close by, they may be able to rescue a firefighter who suffers a critical ailment in rehab.[1]

Other support tasks

Without straying far from their primary function, the rapid intervention team can fulfill a number of other support tasks.

Establishing Equipment Caches. The RIT can stage equipment that may be needed. While the RIT should have their own equipment staging area, many departments like to stage tools, spare SCBA bottles, and other equipment that may be needed at the scene so that they are readily available. One factor in assigning this task is the distance that personnel must travel to obtain the equipment for this cache. One or two hundred feet from the RIT staging area is probably not significant, while two or three blocks probably is. With such extended travel distances, other support personnel must fulfill this task.

Refilling SCBA Bottles. Smaller departments with limited resources may have to assign the RIT to refill SCBA bottles (Figure 9.6). Bigger departments are more likely to have dedicated units to perform this function. The driver of the RIT unit may be able to refill the bottles, as long as he can be freed to perform the RIT's outside safety and communications functions should a rescue be necessary.

Operations Tasks

Finally, the IC may consider assigning some tasks to the RIT, with the full knowledge that he is essentially committing the RIT to the operations branch. This makes them less ready to respond quickly should a sudden, unexpected event occur.

Performing Salvage. Salvage is a lost art in departments throughout the United States, although ironically it is the area where the fire service may gain the greatest public support. Salvage normally is not needed the moment a fire attack begins, but it can be done while or after a fire is being controlled. A major concern is that if salvage is not done soon enough, all may be lost. With a minimum of effort, a few personnel can move furniture and other belongings and then cover them. This may prevent thousands of dollars of property damage. While salvage is certainly important to the wealthy homeowner with many expensive belongings, it can be even more important to homeowners who have little. For them, everything is critical. If no other personnel are available to perform this task, detailing part of the RIT to do it may be a consideration. This will certainly distract the team from their primary function, but it may place them closer to a potential rescue, should one

Figure 9-6 After the fire has been placed under control, the RIT unit serves auxiliary support functions.

be needed. It may also place members of the RIT in the hazard area, where they could be at risk to a sudden, unexpected event.

Relieving Operating Personnel. Personnel operating on the scene may become fatigued just when a critical task must be completed to gain control of an incident. As the operating personnel are relieved of their duties with no replacement personnel immediately available in staging, some RIT personnel might be assigned to complete the critical task. This is another risky situation where the IC must carefully consider the plan of action.

Overhaul is a special task within this overall action. The IC may determine that the situation has been controlled to the point that he can place some or all of the RIT personnel into service doing overhaul. Again, this poses some risks, as sudden, unexpected events can still occur during overhaul. Collapses, rapid increases in fire conditions, and firefighters suffering medical emergencies are certainly not unheard of during over-haul. Again the IC must determine the risks, his available means of dealing with them, and how to deploy his resources. There may be a strong case for utilizing the fresh, RIT personnel for these tasks rather than calling for additional resources that may not be available or that may have an extended response time to the scene.

To reinforce this point, performing these functions places the emergency scene at the greatest risk if something goes wrong. They should be done only if the IC has clearly considered the risks of changing the RIT's role at the incident.

Being prepared to perform this multitude of other tasks both before and after the incident is controlled can make the RIT very versatile. Nonetheless, if the RIT is to perform other duties while standing by to provide primary rescue services for emergency responders, the following conditions must be met:

1. The RIT must stay in close contact with the IC.

2. The integrity of the team must be maintained.

3. The tasks must require relatively little exertion, so as not to significantly diminish the team's physical readiness, should the unexpected occur.

Each department must determine the level of risk it is willing to accept and the legal requirements it must meet to provide an RIT at emergency scenes. The more varied the tasks assigned to the RIT unit, the less likely the unit will be rapidly available should something go wrong. Of course, the likelihood of a sudden, unexpected event diminishes once the incident is controlled and secondary duties begin. This does not mean, though, that sudden, unexpected events are impossible. On the contrary, all proper safety support should be provided for the fire ground to further lessen the likelihood of any sudden, unexpected event.

In some instances, the designated RIT has arrived early at the incident and received an operations sector assignment. For example, the RIT has arrived before the first-due ladder company at a building that must be laddered, forcibly entered, or searched for missing civilians. The IC again must make a risk-based decision whether to hold the RIT in reserve or place it into service. If he chooses the latter, he may reassign the first due ladder company as the RIT upon its arrival on the fire ground. In accordance with OSHA and NFPA standards, however, at least two personnel must remain available at all times to provide firefighter rescue.

Rapid intervention teams can improve safety conditions on the fire ground in numerous ways. The RIT officer, in conjunction with the IC and/or safety officer, must determine which tasks the team can accomplish without excessively fatiguing its members or compromising its high level of readiness. During certain particularly high-risk times on the fire ground, the IC may prefer to keep the RIT on standby, ready to go to work, and doing nothing else. As with a good size-up, the RITs safety actions can literally make the difference between life and death for a fellow firefighter.

Note

For additional information on emergency incident rehabilitation, see "Emergency Incident Rehabilitation," United States Fire Administration (USFA) procedure booklet. This can be ordered online for no charge at www.usfa.fema.gov/usfapubs.

Chapter 10
Accountability

Accountability

An accountability system is the only way to maintain incident control on the fire ground. All area personnel who might operate together on any scene must use the same system. The development of a system must consider all units that might operate on the "big one," not just immediately neighboring departments. An accountability system is as critical for the rapid intervention team as for the personnel operating on the fireground. The RIT must maintain their own accountability and must also have some way of tracking other personnel and units operating at the scene. In turn, the safety officer and/or incident commander must be able to account for all personnel operating as the RIT. How this accountability will be maintained is best determined before the units ever respond to a call.

Accountability Systems

Accountability can be managed in a number of ways. Most common is some type of tag or marker with the member's name on it (Figure 10.1). All manner of tags, key rings, and other devices are used for this function. The tag or marker is kept at a fixed place on the apparatus that personnel ride to the scene. Firefighters responding to the scene report to the apparatus from their department or the command post and "tag-in" there. Typically, this practice is accepted throughout the American fire service. Once these personnel begin to operate on the scene, however, accountability becomes a bit muddled. Some departments use only the accountability system on the apparatus. In these cases, should a sudden, unexpected event occur, someone (likely the IC, his aide, or the safety officer) will have to trudge around to every apparatus operating at the scene to develop accountability for all participating personnel.

One way to account for personnel inside the hazard zone is to develop a tagging system at the ingress points to the building or hazard area (Figure

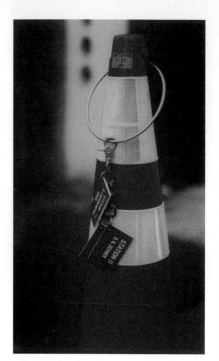

Figure 10.1 Typical firefighter accountability tags. Establishing a system to identify each firefighter working on the fireground is essential.

Figure 10.2 A large cone with a ring, located at the entrance to a fire building, can serve as a highly visible "tag-in" site.

10.2). This can be quite challenging, as tagging in at the entrance to a burning building may be the last thing on the minds of the firefighters making entry. But it may be one of the most important tasks they perform, should they be caught by a sudden, unexpected event. New accountability systems are being developed regularly. Some presently being marketed use bar coding to identify each firefighter. An entry control officer "scans" the firefighters into the hot zone, tracking the sectors to which they are assigned and how long they are in the danger area. The system sounds an alarm when the responders are nearing the end of the SCBA capacity. No matter what system is utilized, whether it be paper, tags, or something more sophisticated, it requires discipline of all personnel on the scene to somehow denote that they have entered the hot zone (Figure 10.3). Ignoring the system, for whatever reason, places personnel operating in the hazard area at high risk. To work effectively, the system must be in place and enforced for all personnel.

A concern with accountability systems is that some departments verify the system and the personnel only when a sudden unexpected event occurs. Since such events are uncommon, most responders are unfamiliar with accountability practices and may not even be keeping track of whom they are working with or for! Once the sudden, unexpected event occurs it is too late to ensure that accountability is being followed properly. Personnel accountability reports should be routinely complete on the fire ground. One way to help make this happen is for the communications sector to provide regular elapsed time reports, or "marks." Section 2-2.6 of NFPA 1561, *Emergency Services Incident Management System* (2000), indicates, "The IC shall be

Figure 10.3 All firefighters operating at an incident must "tag-in" before entering a building.

provided with reports of elapsed time-on-the scene at emergency incidents in 10-minute intervals from the...Communications Center, until reports are terminated by the incident commander." These 10-minute notices give command an opportunity to conduct a fresh size-up of the incident and to determine operational effectiveness. They also indicate the increasing degradation of the fire building's structural integrity, the increasing risk to personnel operating in and around the fire building, and the increasing fatigue for operating personnel. This is the time to gather a personnel accountability report.

Each department must develop standard operating procedures to establish 10 (or 15, or 20) minute marks, and communicating them to the incident commander. The IC, along with the telecommunicators and all personnel operating on the emergency scene, must know what these regular notifications mean. The procedures must be covered in training and practiced and practiced again so that all personnel know how to use them on the fire ground. An effective accountability system will ensure that the RIT knows who is missing, and how many are missing, when a sudden unexpected event occurs. Not having enough resources to search for missing individuals when something goes wrong is bad, but deploying a rapid intervention team into a hazardous area to search for missing individuals who are not really there is awful.

The rapid intervention team also must have a system to maintain their own accountability. Ideally, this system will integrate fully into the overall accountability system in place on the emergency scene. If no system is in place, however, or if the system is ineffective, the RIT must carefully identify their crew and ensure that someone outside the hot zone is tracking them. This function may be assigned to the driver, or "outside man," but it will be much more effective if performed by the safety officer (or incident commander if no safety officer is in place). When the RIT deploys, it likely will be into a high-risk situation. Their accountability must not be ignored. Reporting to the IC initially on arrival also will help ensure that the rescuers will be accounted for.

Ultimately, sector officers are in position to provide the best accountability for their personnel. Unfortunately, some departments do not provide enough supervision, and in others sector officers are uncertain who is reporting to them. Once the RIT is deployed, who do they report to? The sector officer where they are working? The safety officer? The incident commander? This must be worked out in advance. The bottom line is that all emergency responders must know who they are working for, where they are working, and what they are doing. In addition to all of that, officers in charge of sectors must know who is reporting to them and where they are. This is the key to providing optimum accountability on the fire ground.

Mayday procedures

Being lost or missing at an emergency scene is one of the most difficult situations for firefighters to face and come to grips with. They will not want to readily admit being in trouble, but such predicaments indeed occur. When they do, the following rules may just save a life:

1. Do not panic. Stop and think about what is occurring, your location in the building, and how you got there. This may help you find your way out. Following an expansion joint on a floor will generally lead you to a wall.

2. Admit that you are lost and call for help. Follow your department's standard operating procedures for these situations. If your department does not have an SOP, call "Priority" or "Mayday" on your radio, if you have one, and activate the emergency button. Give details of the problem and your location as best as you can determine it. If you move, let command know. If you are not sure exactly where you are, try to reference from a fixed object such as a door or window. A structural steel column on a wall will likely indicate you are on an outside wall.

3. Activate your PASS device. Turn it on to alarm so that it sounds constantly. This will help others find you.

4. Keep your company together. If there is more than one of you, splitting up will not guarantee anything. Discuss your problem and share information. Two minds may discover a solution that one mind would overlook.

5. Follow a hose line or lifeline. Know how hose couplings are arranged. Typically, the male coupling will point away from the apparatus or water discharge point. Following hoses from their female couplings to their male coupling may just bring you outside. This tactic should be practiced in drills, often. Make a spaghetti pile of hose in a dark room, with a female end outside of the building. Personnel should practice following the hose line until they escape.

6. Search for an opening, any opening. A door or window will at least be a point from which you can attract attention. Do not throw your helmet away. If you want to attract attention, try to throw some other object out of the window if you are really in trouble.

7. Conserve your air supply. Long, slow, deep breaths will use up less air than short, rapid breaths. Be keenly aware of backup, buddy breathing methods that you can use with your SCBA. Once the RIT arrives, transfill your cylinder with them.

8. If you are trapped, shine your light upward to attract attention and position your PASS device for maximum effectiveness.

Conclusion

Personnel accountability is a major challenge for the American fire service. It is intimately entwined with the function of the rapid intervention team. If proper accountability is not being followed, the RIT's job will be much more difficult. It may be the RIT's responsibility to perform accountability—a tough job if they do not arrive on the scene early enough in the incident to determine who is working where. Sector officers are a key link in individual accountability, as they must keep track of the personnel assigned to them and be able to report their personnel status to the incident commander on short notice. The means of tracking personnel will improve exponentially as the fire service makes use of various technological advances. In the meantime, we must continue effectively to keep track of operating personnel in more down-to-earth ways. Having systems in place that accomplish this and practicing them frequently will be the only way to ensure that accountability will work the first time, and every time, at major incidents.

On the Web

Additional information on accountability is available at the following web sites:

www.usfa.fema.gov/usfapubs. (United States Fire Administration. Order the free booklet "Personnel Accountability System Technology Assessment Document.")

www.Xtrack.com (Computer-based accountability system manufacturer.)

www.Biosystems.com (Computer-based accountability system manufacturer.)

www.Gracesales.com (Computer-based accountability system manufacturer.)

Conclusion

Every year, responders are injured and killed when things go wrong at emergency scenes. Various organizations have established regulations and standards to address these situations. Rapid intervention is designed to prevent these injuries and deaths, or at least to mitigate the circumstances when they occur. Firefighter deaths are unacceptable, period. Rapid intervention is one element of a progressive safety attitude on the fire ground.

Rapid intervention duties are not to be taken lightly. They require a serious approach and thorough training. Many of these duties simply apply existing equipment and training in a new manner. A great deal of thinking, existing procedures, and resourcefulness is necessary to perform rapid intervention tasks successfully. Search training cannot be emphasized enough. Rescue training specific to the hazards in each department's response area is critical to prepare personnel for rapid intervention. Thinking about rapid intervention assignments will assist firefighters in their own survival training. Implementing a rapid intervention team and training it for these tasks will make every department a better response organization overall. For these reasons, funding for training and equipment for the RIT can generally be absorbed into existing budgets. Furthermore, the benefits of rapid intervention can justify requests for additional funding. While training to rescue the rescuer, firefighters also will become more effective in the skills needed to rescue civilians.

Emergency services organizations must decide in advance how they will use an RIT. Will they simply stand by, ready to perform a rescue when needed, or will they be utilized to perform other critical support functions that will help maintain a high level of safety for all operating personnel?

Rapid intervention preparation and response can be very basic or quite complex. This book proposes a model that will handle the large majority of the risks faced on the fire ground. A two-in/two-out rapid intervention team with a support/logistics person will not only be able to handle most emergencies but will also meet the intent of the appropriate standards for providing backup for the rescuers. If a department chooses to implement this model, it simply must ensure that it knows where to get the resources necessary to handle more challenging situations.

Once a department has justified its need for this type of service and identified the types of hazards to which it likely will respond, it can write procedures and institute training. When it implements the service, the department should incorporate it directly into existing training programs.

Emergency services organizations need to evaluate their ability to respond to sudden, unexpected events at emergency scenes. Do they have a "Plan B" or a "Plan C"? These organizations must be familiar with the standards and regulations that apply to emergency response, and they must develop procedures that enable them to operate within these standards and regulations. Sample procedures and checklists are provided in the appendices.

Each emergency services organization must evaluate all of these factors in order to develop response mechanisms for the potential hazards and available resources in its area. Technology will make fire fighting safer and more precise, but at the same time it will make the risks more complex. Readiness is a must. Together with safety conscious, well trained officers, personnel accountability, and firefighter survival training, rapid intervention can help maintain the well being of those who daily put their lives on the line in every emergency response organization.

Appendices

Bryn Athyn Fire Company Standard Operating Procedure FAST

This procedure outlines actions to be taken when responding to an incident as the "FAST" unit (Firefighter Assist and Search Team).

Purpose

1. The purpose of the Firefighter Assist and Search Team (FAST) is to be immediately available to assist any emergency services personnel who become trapped or are in distress.

Response

1. When dispatched as the FAST unit to working incidents, response will be with Special Service 11, or Engine 11 if the Special Service is not available.

2. All personnel on the unit must be interior qualified firefighters. If 5 interior qualified firefighters are not available, 1 position (rear seat closest to the door) can be filled by an exterior qualified firefighter. Where possible, Firefighter/EMTs should respond with this unit, with a minimum of 2 FF/EMTs aboard. Where possible, the FAST unit should respond with 5 personnel.

Position

1. The officer of the FAST unit shall, upon arrival, report to and remain at the Command Post, unless otherwise directed by the Incident Commander.

2. The entire FAST unit shall remain near the Command Post, within verbal contact distance, at a position from which they can be readily deployed.

Duties

1. The FAST unit shall "stand fast," intact as a unit, ready to take immediate action as directed by the Incident Commander. Avoid

involvement in other fire ground duties unless directly assigned by the IC or Safety Officer.

2. While "standing fast," the unit shall determine the location and availability of portable ladders for future use. They shall also determine the location of EMS personnel and equipment at the scene. The driver or exterior firefighter will be available to position and operate any ladder.

3. The FAST unit shall maintain a state of constant readiness to react rapidly to changing fire ground conditions.

4. Radio messages (Fire and EMS bands) shall be monitored for any indication of members in distress.

Tools and Equipment

1. The officer will monitor radios, and carry a handlight and SCBA.

2. Two firefighters will carry a handlight each, a set of irons, the thermal imaging camera, and SCBA.

3. The driver will maintain the search rope and circular saw.

4. As necessary, the additional firefighter will stage EMS or other needed equipment.

5. Extended duration SCBA will be utilized by all personnel on the FAST unit to provide prolonged service if necessary.

The vehicle driver must position the FAST unit in such a way that its equipment is readily accessible, while not blocking means of access or egress for other units from the incident. The FAST unit is designed as a safety tool for emergency responders. It is important that this service be provided quickly and efficiently.

Sample Rapid Intervention Team Procedure

Appendix 2

1. Purpose The purpose of this Operational Procedure is to define the responsibilities of a RIT, and provide guidelines and procedures for RIT dispatch and RIT operations.

2. Definitions

2.1 Rapid Intervention Team The RIT will be the third due-in ladder company dispatched on a Tactical Box or Box assignment. The RIT will be the fourth due-in ladder company dispatched to a high-rise building. The sole purpose of the RIT is to locate and remove lost or trapped emergency personnel.

2.2 RIT Equipment Firefighters in ladder companies will be assigned RIT tools in addition to the regularly assigned tools. A RIT tool kit will be carried in each Battalion Chief's car. The tool kit contains cutting equipment to augment the tools already carried by the ladder company. The RIT should be prepared to quickly go into service with the following equipment; life belts, hand tools for cutting, rescue rope, portable radios, forcible entry tools such as halligan and ax, saw with multi-purpose blade, rabbit tool, hand lights, wire basket, first-responder bag with resuscitator, additional SCBA and bottles for the emergency breathing support system (EBSS).

3. Responsibilities

3.1 All Members It will be the responsibility of each member to exercise the appropriate control dictated by his/her rank in the implementation of this Operational Procedure. The RIT will be prepared to go into service at a moment's notice. RIT activity at the emergency scene should be consistent with their mission: locating and removing lost or trapped emergency personnel.

3.2 Supervisor, Fire Communications Center (FCC) When notified by the Incident Commander (IC) that a RIT is in service, the FCC Supervisor with notify the appropriate Deputy Chief.

3.3 Incident Commander (IC) The IC will identify the location of the Command Post (CP) and decide when the RIT goes into service.

3.4 Fire Training Academy (FTA) The FTA will be responsible for the training of all members in the proper procedures relating to the implementation of the RIT.

3.5 RIT Company Officer

3.5.1 The RIT Officer will insure that all members of the company are familiar with the duties of the RIT.

3.5.2 The officer or acting officer in charge of the RIT, immediately upon arrival at the incident, will survey the scene. The officer will take full advantage of all information about the structure available in Vital Building Information (VBI) forms and Pre-Fire Plans. The officer will monitor the fire ground radio frequency, note the location of all fire companies, evaluate portable and main ladder placement for firefighter egress, alert the safety officer or IC to unusual structural features and locate all access points to all sectors of the emergency scene.

At roll call the ladder company officer will assign RIT tools to each member of the company.

4. RIT Procedures

4.1 Dispatch

4.1.1 FCC will dispatch an additional ladder company on all tactical box and full box assignments when all companies are placed in service at structure fires, confined space rescue and box assignments to non-structure fires and non-firefighting activities.

4.1.2 The RIT will respond at emergency speed.

4.1.3 If FCC receives a report from the emergency scene that the RIT has been placed in service to locate a lost or trapped firefighter, FCC will dispatch an additional ladder (RIT) at emergency speed.

4.2 Emergency Scene

4.2.1 Upon arrival at the incident the RIT officer will contact the IC and request the location of the CP. The RIT officer will survey the scene while the RIT members assemble RIT equipment. The RIT officer will then report, with his/her team and equipment, to the IC. Once assembled the RIT will remain in the vicinity of the command post, prepared to go into service.

4.2.2 In the event the IC receives a report of a trapped or missing personnel, the IC will order the RIT into service. The RIT will continue search and rescue operations until; the missing or trapped firefighter(s) is (are) located and removed, the RIT is relieved or the RIT is ordered to terminate the operation.

4.2.3 The RIT will operate on the designated emergency scene frequency. The RIT officer will monitor the emergency scene frequency to anticipate any situation where their services will be needed.

Note: This sample procedure is Appendix A in "Rapid Intervention Teams at Structure Fires," an applied research project submitted to the National Fire Academy as part of the Executive Fire Officer Program March 1999, Thomas J. Garrity, M.S., Deputy Chief, Philadelphia Fire Department.

Sample Rapid Intervention Team Checklist

Appendix 3

Size-Up

_____ 1. Building size up (length x width x height).

_____ 2. Building occupancy.

_____ 3. Building construction type:

    _____ Wood frame.

    _____ Heavy timber.

    _____ Ordinary.

    _____ Noncombustible.

    _____ Fire resistive.

_____ 4. Placement of windows, doors, fire escapes, porches, etc.

_____ 5. Potential danger of high-security doors, barred windows, building modifications.

Tactics

_____ 6. Offensive, defensive, defensive-to-offensive

_____ 7. Command operations:

    _____ Check tactics board.

    _____ Check accountability system.

    _____ Communications/incident commanders.

_____ 8. Ladders and truck operations.

_____ 9. Fire ground time vs. progress.

Equipment

_____ 10. Place equipment in staging area (see RIT equipment list).

Other Operations

_____ 11. Check with safety/compare information.

_____ 12. Potential collapse and collapse area.

_____ 13. Relocate or add more RIT.

_____ 14. Location of EMS unit.

Note: This sample checklist is Appendix C in "The Development and Use of Rapid Intervention Teams for the Chelmsford, MA, Fire Department," an applied research project submitted to the National Fire Academy as part of the Executive Fire Officer Program, September 1999, John E. Parow, Fire Chief, Chelmsford Fire Department, Chelmsford, MA. Referenced from Rick Kolomay and Bob Hoff, "Saving Our Own: The Rapid Intervention Team Checklist," *Fire Engineering,* January 1998.

"RIC" Crew Assignment

SAMPLE CHECKLIST

____ Report to IC.

____ Turn in "PATs" to IC.

____ Tools & Equipment Required:

    ____ Full Turnout Gear w/ SCBA

    ____ Portable Radio

    ____ 2 Spare SCBA Bottles

    ____ Forcible Entry Tools

    ____ Power Saw

    ____ Search Rope

    ____ Rescue Lifeline

    ____ SFC Highband Radio, F6 (Located on SFC apparatus, see driver of apparatus)

____ Size-up Scene

    ____ Entry & Egress Points

    ____ Fire/Hot Zone Location

    ____ Firefighting/Rescue Operations

    ____ Hazards in & around Area

    ____ Additional Equipment Resources (ladders, attack lines, rescue equipment, etc.)

_____ Establish Secondary Egress Route.
(Place a ground ladder and raise to fire floor, floor above fire or roof.)

_____ For Commercial Operation—(Establish ladder apparatus placement with stabilizers set-up and ready for immediate service.)

_____ Assist IC. (Emergency Communications & additional Hazard Assessment)

_____ Stay together and be ready at all times!

_____ Released for reassignment only by IC.

(Courtesy of Southampton Fire Co., Bucks County, Pennsylvania)

Drill Ideas/Training

DRILL I

Hide a PASS device that is activated in a room. Send firefighters in to search for it. The level of difficulty of the room and the level of smoke or mask blackout provided can depend upon the skill level of the firefighters involved.

DRILL II

Make a spaghetti pile of fire hose (2 1/2-inch or smaller) in a room or area. Place the female end outside, or connect to a piece of apparatus. With smoke or mask blackout, have firefighters find the hose, then find their way out of the building.

DRILL III

During any regular drill, have a firefighter "go down." Allow the rapid intervention team, if there is one, to go into action.

RIT SCENARIOS

Tabletop these situations with your department. You are the chief in charge of the following situations:

1. You are the commanding officer of an attached garage fire in a single family dwelling. There are heavy fire conditions in the garage, with fire spreading into the attic of the dwelling. Three engines and a ladder company have been in service for 15 minutes, although interior officers are reporting limited progress (man-power is light). Firefighters are on the roof of the home opening up. A BLS ambulance is staged near the command post. As the interior sector commander comes out to the command post to give you a report, the roof collapses, taking two firefighters into the attic.

What are your priorities?

What tactics will be needed?

What resources will be needed? What communications are given via radio? If you were the FAST officer, how would you set up at this scene, and how would you react to this sudden, unexpected event?

2.   You are the commanding officer at a 50-by-100-foot, 4-story manufacturing facility of ordinary construction. Heavy smoke is showing from the third floor, with medium smoke throughout the fourth floor. Two engines are conducting fire attack with hand lines, two engines focusing on water supply for the attack engines, one ladder company is attempting horizontal ventilation, but having difficulty with bricked-up and metal covered windows. The second ladder company is in service opening the roof. An ALS ambulance is staged near the command post. The rescue company is en route with a manpower contingent of four. You have just requested an additional engine and ladder to stage in the parking lot across the street when a captain stumbles out of the building. A flashover has occurred on the third floor. Three of his firefighters are missing. They were operating in the building for at least 15 minutes when this occurred.

What are your priorities?

What tactics will be needed?

What resources will be needed? What communications are given via radio? If you were the FAST officer, how would you set up at this scene, and how would you react to this sudden, unexpected event?

Note: Tabletop scenarios paraphrased from "Firefighter Assistance and Search Team," a presentation by Dr. Jim Cline, EDCON Associates, 1996.

Selected Bibliography

Baker, Michael and Joseph Ross. "The 3 Ws of Saving Our Own." *Firehouse*, July 2000, pp. 82–84.

Carlson, Gene (ed.). *Forcible Entry*, 7th ed. Stillwater, Oklahoma: Fire Protection Publications, 1987.

Cline, Jim. "Rapid Intervention Companies: The Firefighter's Life Insurance." *Fire Engineering*, June 1995, pp. 67–68.

Cline, Jim. "Firefighter Assistance and Search Team." Presentation, developed 1996.

Cobb, Robert. "Rapid Intervention Teams: They May Be Your Only Chance." *Firehouse*, July 1996, pp. 54–57.

Cobb, Robert. "Rapid Intervention Teams: A Fireground Safety Factor." *Firehouse*, May 1998, pp. 52–56.

Coleman, John and Rick Lasky. "Managing the Mayday." *Fire Engineering*, January 2000, pp. 51–62.

Crawford, James. "Rapid Intervention Teams: Are You Prepared for the Search?" *Firehouse*, April 1999, pp. 50–54.

Donahue, Art. "RIT Rope Drag." *Fire Engineering*, February 1999, p. 14.

Dubé, Robert. "Rescue Rope for Rapid Intervention Teams." *Fire Engineering*, July 2000, p. 14.

Dugan, Michael. "How to Use a Rapid Intervention Team." *Firehouse*, July 1996, pp. 58–62.

Dunn, Vincent. "Disorientation: A Firefighter Killer." *Firehouse*, November 1999, pp. 18–21.

Fire Department Safety Officers Association. "Rapid Intervention Teams: If You Don't Have One, Get One." *Safety-Gram*, January 2000.

Fredericks, Andrew. "Engine Company Support of RIT/FAST Operations." *Fire Engineering*, April 1999, pp. 79–96.

Garrity, Thomas. *Rapid Intervention Teams at Structure Fires*. Emmitsburg, Maryland: National Fire Academy, 1999.

Hall, Richard and Barbara Adams, eds. *Essentials of Firefighting*, 4th ed. Stillwater, Oklahoma: Fire Protection Publications, 1998.

Howard, George. "The Search Rope." *Fire Engineering*, January 1992, pp. 12–15.

Jakubowski, Greg. "Better Get in FAST: The Firefighter Assist and Search Team." *Firefighter's News*, March 1996, pp. 30–32.

Jakubowski, Greg. "On The Mark: Four Smart Reasons for Your Service to Understand and Implement the 20-Minute Mark." *Fire Rescue*, October 2000, pp. 46–49.

Jakubowski, Greg and Mike Morton. *Rapid Intervention Teams*. Program developed for the Indiana Fire Instructors' Association, 1998.

Jonas, John. "The FAST Unit: What You Should Consider." *WNYF*, June 1999, pp. 13–15.

Kolomay, Rick and Bob Hoff. "Saving Our Own: The Rapid Intervention Team Checklist." *Fire Engineering*, January 1998, pp. 12–19.

Lasky, Rick. "Saving Our Own: The Rapid Intervention Team Officer." *Fire Engineering*, July 1997, pp. 17–20.

Lasky, Rick. "Saving Our Own: Designing a Firefighter Survival Training Aid." *Fire Engineering*, May 1998, pp. 10–13.

Lasky, Rick and Ray Hoff. "Saving Our Own: The Firefighter Who Has Fallen through the Floor." *Fire Engineering*, March 1998, pp. 12–18.

Lasky, Rick and Rick Kolomay. "Saving Our Own: Approaching a Downed Firefighter." *Fire Engineering*, September 1997, pp. 14–18.

Lasky, Rick and Sal Marchese. "Saving Our Own: Removing a Downed Firefighter from a Tight Room and Window." *Fire Engineering*, April 1998, pp. 12–20.

Lasky, Rick and Tom Shervino. "Saving Our Own: Moving the Downed Firefighter up a Stairwell." *Fire Engineering*, December 1997, pp. 14–18.

Lund, Pete. "Tactical Considerations for the Rapid Intervention Team." *Firehouse*, September 1999, p. 86.

Manning, William, ed. "Roundtable: Rapid Intervention Teams." *Fire Engineering*, July 2000, pp. 20–38.

McCormack, Jim. "Rapid Intervention: Emergency Air Supply." *Fire Engineering*, July 2000, pp. 14–18.

McLees, Mark. "The Rapid Intervention Rope Bag." *Firehouse*, April 2000, pp. 40–42.

Morton, Mike. "FAST Teams: Prepare an Action Plan Using a Two-Team Method." *Fire Rescue*, January 1999, pp. 75–76.

Norman, John. "Rapid Intervention Techniques." *Firehouse*, November 1997.

Norman, John and John Jonas. *Training Bulletin: Unconscious Firefighter Removal*. New York City Fire Department Training Document, undated.

Olsen, Jay. "AWARE: A Life-Saving Plan for Rescuing Trapped Firefighters." *Fire Engineering*, December 1998, pp. 52–55.

Ordway, Kieran. "Fallen Firefighter Drag Rescue." *Fire Engineering*, August 1999, pp. 54–56.

Parrow, John. *The Development and Use of Rapid Intervention Teams for the Chelmsford, MA, Fire Department*. Emmitsburg, Maryland: National Fire Academy, 1999.

Robertson, Michael. "Safety in Numbers." *Fire Chief*, May 2000, pp. 54–58.

United States Fire Administration. *Firefighter Fatalities in the United States*, 1990–1999 eds. Federal Emergency Management Agency, 1990–1999.

Additional information on firefighter deaths can be found at the following web sites:

www.usfa.fema.gov/usfapubs (United States Fire Administration)

www.usfa.fema.gov/ffmem (United States Fire Administration)

www.cdc.gov/niosh/firehome.html (National Institute for Occupational Safety and Health)

www.nfpa.org (National Fire Protection Association)—has summary information on firefighter deaths.

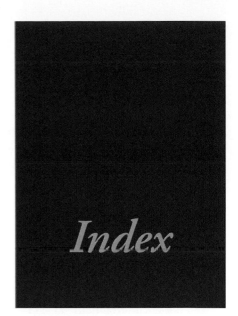

Index

two-rope method of through-the-floor evolutions 106, 107
two-team RITs
    described 73–75
    entry team of 73–74, 76–78, 80–85
    outside team of 74
    second (support) team of 74–75, 85, 119–139, 137–138

U

"Unconscious Firefighter Removal" *112, 139*
United States Fire Administration (USFA)
    firefighter fatality information 175
    ordering booklets online *148*
    web site about accountability *155*
urban search and rescue (USAR) 68
utilities, controlling 145

V

vehicles providing RIT services 45–49
ventilation, casualties caused by inadequate 11, 14
victims
    assessing 80–84
    freeing entangled 93
    locating 78, 79–80
    moving
        at crawling operations 95, 96–97, 105
        down ladders 122–132
        down stairs 104–105
        drags used for 94–98
        fireman's carry 98–99
        pulley systems for 99–100
        up stairs 100–104
        when trapped below area of operation 105–108
        into windows 119–122
    thermal imaging cameras not showing 78

W

walls, breaching 132–137
water supplies. *See also* hose lines
    backup, failure to establish 16
    controlling 145
    ensuring availability of 138
webbing drags for moving victims 94–97, 96–97
web sites
    accountability information *155*
    firefighter fatality information 175
    for OSHA *39*
    for USFA *155*
wildland/brush fire incidents 68
windows
    ladder slide method for window rescues 131–132
    moving victims into 119–122
    rescues using ropes 125–132
witness accounts, searches relying on 80
wooden walls, breaching 134, 135